SAS® Functions
by Example
Second Edition

Ron Cody

**THE
POWER
TO KNOW.**

The correct bibliographic citation for this manual is as follows: Cody, Ron. 2010. *SAS® Functions by Example, Second Edition.* Cary, NC: SAS Institute Inc.

SAS® Functions by Example, Second Edition

Copyright © 2010, SAS Institute Inc., Cary, NC, USA

ISBN 978-1-60764-364-7 (electronic book)
ISBN 978-1-60764-340-1

SAS Institute Inc., SAS Campus Drive, Cary, North Carolina 27513.

1st printing, March 2010
2nd printing, August 2011

SAS® Publishing provides a complete selection of books and electronic products to help customers use SAS software to its fullest potential. For more information about our e-books, e-learning products, CDs, and hard-copy books, visit the SAS Publishing Web site at **support.sas.com/publishing** or call 1-800-727-3228.

SAS® and all other SAS Institute Inc. product or service names are registered trademarks or trademarks of SAS Institute Inc. in the USA and other countries. ® indicates USA registration.

Other brand and product names are registered trademarks or trademarks of their respective companies.

Contents

List of Programs

Chapter 1

Chapter 2

Chapter 7

Chapter 8

Chapter 9

Chapter 10

Chapter 16

Preface to the Second Edition

It is hard to believe that it's been over six years since the first edition of this book was published. Although the major changes in SAS functions came with SAS®9, many new functions have been added to SAS since then, as well as additional capability added to old functions (such as new arguments added to existing functions).

In this edition, I removed the chapter on SAS regular expressions and updated the chapter on Perl regular expressions. I also added a new chapter on the two CALL routines that sort values within an observation (CALL SORTN and CALL SORTC). As you will see in the examples of these CALL routines, they can be extremely useful. In addition, features new to SAS 9.2 are noted.

Indeed, SAS has so many functions that it is hard for you to decide which functions are indispensable and which ones you can do without. When I teach my Functions course, I like to list what I call the "blockbuster" functions—those that change the way we program. Here are a few of my favorites:

The MISSING function takes either a character or numeric argument and returns a value of True if the argument is a missing value, and False otherwise. I no longer write statements such as

```
if Age = . then Age_group = .;
```

Instead, I write

```
if missing(Age) then Age_group = .;
```

or even,

```
if missing(Age) then call missing (Age_group);
```

Several others come to mind: The COMPRESS function with modifiers such as 'kd' is also one of my favorites (take a look at some of the examples of this function in Chapter 1). The SAMEDAY argument to the INTNX function adds huge functionality to this function (along with the new

HOLIDAY function). I also recommend to all my students that they use the three FIND functions (FIND, FINDC, and FINDW) instead of the older INDEX functions. I used to think the concatenation functions were not all that useful; I have changed my mind about that. I use them all the time.

I hope you find your personal favorites in this book.

Ron Cody
Winter, 2010

Preface to the First Edition

SAS functions provide some of the real power of SAS. This book covers almost two hundred of the most common and useful functions and call routines—I did not attempt to describe every one. Unlike SAS language manuals or other reference materials where SAS functions are merely described, this book shows you the functions and one or more examples of how they can be used. These examples are complete working programs. Sometimes the examples demonstrate a common usage of the function; at other times, the use is non-traditional, but still very practical. Some of the examples show how several functions can be used in combination to produce a desired result.

This book contains useful information for veteran SAS users as well. For example, SAS now supports Perl regular expressions. In addition, new arguments have been added to old functions (such as COMPRESS and SCAN) that enhance their usefulness.

One of the strengths of SAS is its ability to manipulate character data. You will find extensive examples of character manipulation in this book. For example, there are complete programs to perform "fuzzy matching" using a combination of functions, including the SPEDIS (spelling distance) function.

For those who are statistically inclined, there are examples of random number functions performing Monte Carlo simulations, including the use of the SAS Output Delivery System (ODS) to capture values from a procedure to a SAS data set.

I have decided to stay with the same function categories described in the latest version of the *SAS OnlineDoc*. Because some of these functions have other important uses besides the obvious one described by its category, I have provided two lists to allow you to look up a function by name, and by program. There is also a traditional index that includes functions and tasks. Note: The wording of arguments in this book might differ from the wording of arguments in the *SAS OnlineDoc*.

Besides providing a nearly complete description of the Base SAS functions, I have provided programs that I hope you will find useful in your daily programming tasks. Several of the more useful examples here are also presented as SAS macros. All the programs in this book can be downloaded from the author page for this book, located at **support.sas.com/authors**. The programs can be used as is, or modified to fit your particular application.

By the way, this is the first book I have ever written while on sabbatical. In fact, this is the first sabbatical I have ever had. I have truly enjoyed writing this book, even though I was "forced" to embark on research trips to Virgin Gorda and Italy (a villa in the Chianti region). After several bottles of excellent wine, I was able to conjure up some interesting examples to add to this book.

Acknowledgments

I don't suppose many folks read the Acknowledgments—and that is too bad. Although this book has only one name on the cover, mine, it really takes a team to produce a book. To start with, I had a team of reviewers, some from SAS, and others who are friends and colleagues not affiliated with SAS. Some members of this team were responsible for only one or two chapters—others read the book from cover to cover. Reviewing a book, and doing a good job of it, is a huge job. I'm not sure why all these people volunteered to do this, but I am extremely grateful to them. As a matter of fact, several of the reviewers have reviewed several of my books—and they keep coming back for more! So, it is with sincere thanks that I acknowledge the following members of my review team:

Mike Zdeb, Paul Grant, Jason Secosky, Kurt Jones, Scott McElroy, Charley Mullin, Kathy Passarella, Kent Reeve, Russ Tyndall, and Sally Walczak

Next, no book project could go forward without a good editor, and I have a great one. John West is amazing. He is always positive and makes my job so much easier. He is also assisted by many other experts, namely:

Joel Byrd, copyeditor
Candy Farrell, technical publishing specialist
Shelly Goodin, marketing
Stacey Hamilton, marketing
Patrice Cherry, designer - front cover
Jennifer Dilley, designer - back cover
Mary Beth Steinbach, managing editor

Finally, as many fiction authors like to do, I would also like to acknowledge my wife's support in my writing project. Thanks, Jan.

Introduction: A Brief Discussion of SAS Functions and Call Routines

If you have written any SAS program, you probably used several functions and have an intuitive idea of what functions do. SAS functions perform a computation or manipulation of a value and return a single value. For example, the MONTH function computes the month of the year from a SAS date; the ROUND function can round numeric values; the TODAY function returns the current date.

SAS functions take the form of a function name, followed by a set of parentheses. In these parentheses are usually one or more arguments. (Certain functions such as TODAY take no arguments, but you need to follow the function name with an opening parenthesis and a closing parenthesis anyway to tell SAS that you are referring to the TODAY function, not to a variable called TODAY.) These arguments provide information that the function needs to return a result. For example, the statement `DAY_OF_WEEK = WEEKDAY(DATE);` computes the day of the week from a SAS date and assigns it to the variable DAY_OF_WEEK. This function, as with all other SAS functions, returns a single value that is either assigned to a variable or used in a SAS expression.

Finally, the arguments to most SAS functions can be variable names, constants (if a character argument is needed, the constant must be in single or double quotation marks), or expressions (that may also contain other functions).

CALL routines have some similarity to SAS functions because they often perform similar operations. However, there are some important differences. CALL routines are not used in assignment statements. Instead, they stand alone as SAS statements. Multiple arguments in a CALL routine can be assigned new values by the routine. For example, the statement `CALL SCAN(string, n, position, length);` returns the position and the length of the *n*th "word" in the string. The SCAN function, on the other hand, returns only a single value—the *n*th word in the string. Indeed, the main reason for using a CALL routine instead of a function is to obtain more than one value in a single statement.

Functions and CALL routines in this book are arranged by category and topic. For example, in Chapter 1, "Character Functions," there are topics such as "Functions That Search for Characters" or "Functions That Remove Leading and Trailing Blanks from Strings." This author hopes that this arrangement (as opposed to alphabetical order) will make it easier for you to find the function you need quickly.

C h a p t e r   1

Character Functions

Introduction

A major strength of SAS is its ability to work with character data. The SAS character functions are essential to this. The collection of functions and CALL routines in this chapter enable you to do extensive manipulation on all sorts of character data.

With the introduction of SAS®9, the number of functions increased greatly (especially character functions). In addition, many of the new character functions included arguments that refer to character classes (such as all digits or all punctuation), making them even more powerful.

After reading this chapter, you will also want to review the next chapter on Perl regular expressions, another way to process character data.

Before delving into the realm of character functions, it is important to understand how SAS stores character data and how the length of character variables gets assigned.

Storage Length for Character Variables

It is in the compile stage of the DATA step that SAS variables are determined to be character or numeric, that the storage lengths of SAS character variables are determined, and that the descriptor portion of the SAS data set is written. The following program will help you to understand how character storage lengths are determined:

Program 1.1: How SAS determines storage lengths of character variables

```
data example1;
   input Group $
      @10 String $3.;
   Left  = 'x    ';  *x and 4 blanks;
   Right = '    x';  *4 blanks and x;
   Sub = substr(Group,1,2);
   Rep = repeat(Group,1);
datalines;
ABCDEFGH 123
XXX         4
Y           5
;
```

Explanation

The purpose of this program is not to demonstrate SAS character functions. That is why the functions in this program are not printed in **bold** as they are in all the other programs in this book. Let's look at each of the character variables created in this DATA step. To see the storage length for each of the variables in data set EXAMPLE1, let's run PROC CONTENTS. Here is the program:

Program 1.2: Running PROC CONTENTS to determine storage lengths

```
title "PROC CONTENTS for Data Set EXAMPLE1";
proc contents data=example1 varnum;
run;
```

The VARNUM option requests the variables to be in the order that they appear in the SAS data set, rather than the default, alphabetical order. The output is shown next:

```
Variables in Creation Order

#    Variable    Type    Len

1    Group       Char      8
2    String      Char      3
3    Left        Char      5
4    Right       Char      5
5    Sub         Char      8
6    Rep         Char    200
```

First, GROUP is read using list input. No informat is used, so SAS will give the variable the default length of 8. Since STRING is read with an informat, the length is set to the informat width of 3. LEFT and RIGHT are both created with an assignment statement. Therefore the length of these two variables is equal to the number of characters between the quotes. In this program, single quotes are used to create LEFT and RIGHT. The program would produce the same result if double quotes were used. In most cases, you can use either single or double quotes to create a character constant. Note that if a variable appears several times in a DATA step, its length is determined by the **first** reference to that variable.

For example, beginning SAS programmers often get in trouble with statements such as:

```
if Sex = 1 then Gender = 'Male';
else if Sex = 2 then Gender = 'Female';
```

The length of GENDER in the two previous lines is 4, since the statement in which the variable first appears defines its length.

There are several ways to make sure a character variable is assigned the proper length. Probably the best way is to use a LENGTH statement. So, if you precede the two previous lines with the statement:

```
length Gender $ 6;
```

the length of GENDER will be 6, not 4. Some lazy programmers will "cheat" by adding two blanks after MALE in the assignment statement (me, never!). Another trick is to place the line for FEMALE first.

So, continuing on to the last two variables. You see a length of 8 for the variable SUB. As you will see later in this chapter, the SUBSTR (substring) function can extract some or all of one string and assign the result to a new variable. Since SAS has to determine variable lengths in the compile stage and since the SUBSTR arguments that define the starting point and the length of the substring could possibly be determined in the execution stage (from data values, for example), SAS does the logical thing: it gives the variable defined by the SUBSTR function the longest length it could possibly need—the length of the string from which you are taking the substring.

Finally, the variable REP is created by using the REPEAT function. As you will find out later in this chapter, the REPEAT function takes a string and repeats it as many times as directed by the second argument to the function. Using the same logic as the SUBSTR function, since the length of REP is determined in the compile stage and since the number of repetitions could vary, SAS gives it a default length of 200. A note of historical interest: Prior to SAS 7, the maximum length of character variables was 200. With the coming of SAS 7, the maximum length of character variables was increased to 32,767. SAS made a very wise decision to leave the default length at 200 for situations such as the REPEAT function described here. The take-home message is that you should always be sure that you know the storage lengths of your character variables.

Functions That Change the Case of Characters

Two old functions, UPCASE and LOWCASE, change the case of characters. As of SAS®9, a new function, PROPCASE (proper case), capitalizes the first letter of each word.

Function: **UPCASE**

Purpose: To change all letters to uppercase.

Note: The corresponding function LOWCASE changes uppercase to lowercase.

Syntax: UPCASE(*character-value*)

character-value is any SAS character value.

If a length has not been previously assigned, the length of the resulting variable will be the length of the argument.

Examples

For these examples, CHAR = "ABCxyz".

Function	Returns
UPCASE(CHAR)	"ABCXYZ"
UPCASE("al%m?")	"Al%M?"

Program 1.3: Changing lowercase to uppercase for all character variables in a data set

```
***Primary function: UPCASE
***Other function: DIM;

data mixed;
   length a b c d e $ 1;
   input a b c d e x y;
datalines;
M f P p D 1 2
m f m F M 3 4
;
data upper;
   set mixed;
   array all_c[*] _character_;
   do i = 1 to dim(all_c);
      all_c[i] = upcase(all_c[i]);
   end;
   drop i;
run;
```

```
title 'Listing of Data Set UPPER';
proc print data=upper noobs;
run;
```

Explanation

Remember that uppercase and lowercase values are represented by different internal codes, so if you are testing for a value such as Y for a variable and the actual value is y, you will not get a match. Therefore, it is often useful to convert all character values to either uppercase or lowercase before doing your logical comparisons. In this program, _CHARACTER_ is used in the array statement to represent all the character variables in the data set MIXED. Inspection of the following listing verifies that all lowercase values were changed to uppercase.

```
Listing of Data Set UPPER

a    b    c    d    e    x    y

M    F    P    P    D    1    2
M    F    M    F    M    3    4
```

Function: **LOWCASE**

Purpose: To change all letters to lowercase.

Syntax: **LOWCASE(*character-value*)**

character-value is any SAS character value.

Note: The corresponding function UPCASE changes lowercase to uppercase.

If a length has not been previously assigned, the length of the resulting variable will be the length of the argument.

Examples

For these examples, CHAR = "ABCxyz".

Function	Returns
LOWCASE(CHAR)	"abcxyz"
LOWCASE("A1%M?")	"a1%m?"

Program 1.4: Program to demonstrate the LOWCASE function

```
***Primary function: LOWCASE;

data wt;
    Input ID : $3. Unit : $1. Weight;
    Unit = lowcase(Unit);
    if Unit eq 'k' then Wt_lbs = 2.2*Weight;
    else if Unit = 'l' then Wt_lbs = Weight;
datalines;
001 k 100
002 K 101
003 L 201
004 l 166
;
title "Listing of Data Set WT";
proc print data=wt noobs;
run;
```

Explanation

This program demonstrates a common problem—dealing with character data in mixed case. Here you see that the values of the variable Unit are sometimes in uppercase and sometimes in lowercase. A simple way to solve this problem is to use the LOWCASE function to ensure that all values of Unit will be in lowercase.

You could also have used the UPCASE function to convert all the Unit values to uppercase.

Note: If you have mixed case data, you can use the $UPCASE informat to convert the values to uppercase as they are being read. There is no corresponding $LOWCASE informat.

The remainder of the program is straightforward: after you convert all the values of Unit to lowercase, you only have to check for lowercase k's and l's to determine whether you need to multiply the Weight value by 2.2 or not. The following listing shows that this program worked as desired (the variables Unit and Weight would normally be dropped in a program like this):

```
Listing of Data Set WT

ID      Unit    Weight    Wt_lbs

001      k       100      220.0
002      k       101      222.2
003      l       201      201.0
004      l       166      166.0
```

Function: **PROPCASE**

Purpose: To capitalize the first letter of each word in a string.

Syntax: PROPCASE (*character-value* <,*delimiters*>)

character-value is any SAS character value.

Delimiters An optional list of word delimiters. The default delimiters are blank, forward slash, hyphen, open parenthesis, period, and tab.

If a length has not been previously assigned, the length of the resulting variable will be the length of the argument.

Examples

For these examples, CHAR = "ABCxyz".

Function	Returns
PROPCASE (CHAR)	"Abcxyz"
PROPCASE ("al%m?")	"Al%m?"
PROPCASE ("mr. george w. bush")	"Mr. George W. Bush"
PROPCASE ("l'oeuf")	"L'oeuf" ("the egg" in French)
PROPCASE ("l'oeuf"," ' ")	"L'Oeuf"

Note: It seems useful to use both blanks and single quotes as delimiters since you would like names such as D'Allesandro to have an uppercase letter following the single quote.

Program 1.5: Capitalizing the first letter of each word in a string

```
***Primary function: PROPCASE;

data proper;
   input Name $60.;
   Name = propcase(Name);
datalines;
ronald cODy
THomaS eDISON
albert einstein
;
```

```
title "Listing of Data Set PROPER";
proc print data=proper noobs;
run;
```

Explanation

In this program, you use the PROPCASE function to capitalize the first letter of the first and last names. Here is the listing:

```
Listing of Data Set PROPER

     Name

Ronald Cody
Thomas Edison
Albert Einstein
```

Program 1.6: Program to capitalize the first letter of each word in a string

```
***First and last name are two separate variables;

data proper;
   informat First Last $30.;
   input First Last;
   length Name $ 60;
   Name = catx(' ', First, Last);
   Name = propcase(Name);
datalines;
ronald cODy
THomaS eDISON
albert einstein
;
title "Listing of Data Set PROPER";
proc print data=proper noobs;
run;
```

Explanation

In this program, the first and last names are stored in separate variables and you want a new variable (NAME) that contains both first and last name (in proper case).

The CATX function is used to concatenate the first and last name with a blank as the separator character. The PROPCASE function is then used to capitalize each name. The listing is identical to the preceding listing.

Functions That Remove Characters from Strings

COMPBL (compress blanks) can replace multiple blanks with a single blank. The
COMPRESS function can remove not only blanks, but also any characters you specify from
a string.

Function: COMPBL

Purpose: To replace all occurrences of two or more blanks with a single blank
character. This is particularly useful for standardizing addresses and names
where multiple blanks may have been entered.

Syntax: COMPBL(*character-value*)

character-value is any SAS character value.

If a length has not been previously assigned, the length of the resulting
variable will be the length of the argument.

Example

For these examples, CHAR = "A C   XYZ".

Function	Returns
COMPBL(CHAR)	"A C XYZ"
COMPBL("X Y Z LAST")	"X Y Z LAST"

Program 1.7: Using the COMPBL function to convert multiple blanks to a single blank

```
***Primary function: COMPBL;

data squeeze;
   input #1 @1  Name    $20.
         #2 @1  Address $30.
         #3 @1  City    $15.
            @20 State    $2.
            @25 Zip      $5.;
   Name = compbl(Name);
   Address = compbl(Address);
   City = compbl(City);
datalines;
Ron Cody
```

```
89 Lazy Brook Road
Flemington        NJ    08822
Bill      Brown
28  Cathy   Street
North   City      NY    11518
;
title 'Listing of Data Set SQUEEZE';
proc print data=squeeze;
   id Name;
   var Address City State Zip;
run;
```

Explanation

Each line of the address was passed through the COMPBL function to replace any sequence of two or more blanks to a single blank. Here is a listing of data set SQUEEZE:

```
Listing of Data Set SQUEEZE

  Name           Address          City       State    Zip

Ron Cody     89 Lazy Brook Road   Flemington    NJ    08822
Bill Brown   28 Cathy Street      North City    NY    11518
```

Function: COMPRESS

Purpose: To remove specified characters from a character value. When used with the k modifier, compress can be used to keep characters (such as digits) in a character value and remove everything else.

Syntax: COMPRESS(*character-value* <,'*compress-list*'> <,'*modifiers*'>)

character-value is any SAS character value.

compress-list is an optional list of the characters you want to remove. If this argument is omitted, the default character to be removed is a blank. If you include a list of characters to remove, only those characters will be removed (unless the 'k' modifier is used). If a blank is not included in the list, blanks will not be removed.

modifiers is an optional list of modifiers that refer to character classes. For example, the a modifier refers to all uppercase and lowercase characters. The following are some of the more useful modifiers:

a	include all uppercase and lowercase letters.
d	include all digits.
s	include all space characters (blanks, horizontal and vertical tabs, carriage return, linefeed, and formfeed).
i	ignore the case of characters.
k	keep the characters listed in the *compress-list* or referenced by the modifiers instead of removing them.
p	include all punctuation characters.

Note that the modifiers can be in uppercase or lowercase.

If a length has not been previously assigned, the length of the resulting variable will be the length of the first argument.

Examples

In the following examples, CHAR = "A C123XYZ".

Function	Returns
COMPRESS("A C XYZ")	"ACXYZ"
COMPRESS("(908) 777-1234"," (-)")	"9087771234"
COMPRESS(CHAR,,'as')	"123"
COMPRESS(CHAR,"0123456789")	"A CXYZ"
COMPRESS(CHAR,,'d')	"A CXYZ"
COMPRESS(CHAR,"ABC",'d')	" XYZ"
COMPRESS(CHAR,,'kd')	"123"

Program 1.8: Removing dashes and parentheses from phone numbers

```
   ***Primary function: COMPRESS;

data phone_number;
   input Phone $ 1-15;
   Phone1 = compress(Phone);
   Phone2 = compress(Phone,'(-) ');
   Phone3 = compress(Phone,,'kd');
datalines;
(908)235-4490
(201) 555-77 99
;
```

```
title 'Listing of Data Set PHONE_NUMBER';
proc print data=phone_number;
run;
```

Explanation

For the variable PHONE1, the second argument is omitted from the COMPRESS function; therefore, only blanks are removed. For PHONE2, left and right parentheses, dashes, and blanks are listed in the second argument, so all of these characters are removed from the character value. Finally, for PHONE3, the two modifiers k (keep) and d (digits) are used to keep all the digits in the PHONE value and remove everything else. Note that there are two commas following the first argument. If you used only one comma here, SAS would think you wanted to remove k's and d's from the PHONE value. Even though the results for PHONE2 and PHONE3 are identical, it is usually better to specify what you want to keep rather than what you want to remove. Why? Suppose your original PHONE value contained some unwanted character such as a punctuation mark or even a non-printing character. Using the k and d modifiers ensures that you wind up with only digits in the result.

The following listing shows the output:

```
Listing of Data Set PHONE_NUMBER

Obs        Phone            Phone1           Phone2        Phone3

 1     (908)235-4490     (908)235-4490     9082354490    9082354490
 2     (201) 555-77 99   (201)555-7799     2015557799    2015557799
```

Converting Social Security Numbers to Numeric Form

Here is another example where the COMPRESS function makes it easy to convert a standard Social Security number, including the dashes, to a numeric value.

Program 1.9: Converting Social Security numbers from character to numeric

```
***Primary function: COMPRESS
***Other function:  INPUT;

data social;
   input @1 SS_char $11.
         @1 Mike_Zdeb comma11.;
   SS_numeric = input(compress(SS_char,'-'),9.);
   SS_formatted = SS_numeric;
   format SS_formatted ssn.;
```

```
datalines;
123-45-6789
001-11-1111
;
title "Listing of Data Set SOCIAL";
proc print data=social noobs;
run;
```

Explanation

The COMPRESS function is used to remove the dashes from the Social Security number and the INPUT function does the character-to-numeric conversion.

It should be noted here that the Social Security number, including dashes, can be read directly into a numeric variable using the comma11. informat. This trick was brought to light by Mike Zdeb in a NESUG workshop in Buffalo in the Fall of 2002. Here, the variable SS_FORMATTED is set equal to the variable SS_NUMERIC so that you can see the effect of adding the SSN. format. (Note that SSN. is equivalent to SSN11.) This format prints numeric values with leading zeros and dashes in the proper places, as you can see in the following listing:

```
Listing of Data Set SOCIAL

  SS_char      Mike_Zdeb     SS_numeric     SS_formatted

123-45-6789    123456789      123456789     123-45-6789
001-11-1111     1111111        1111111      001-11-1111
```

Counting the Number of Digits in a Character String

This program computes the number of digits in a string by a novel method. It uses the COMPRESS function to remove all digits from the string and then subtracts the resulting length from the original length for the computation.

Program 1.10: Counting the number of digits in a string

```
***Primary functions: COMPRESS, LENGTHN;

data count;
   input String $20.;
   Only_letters = compress(String,,'d');
   Num_numerals = lengthn(String) - lengthn(Only_letters);
datalines;
ABC123XYZ
```

```
XXXXX
12345
1234X
;
title "Listing of Data Set COUNT";
proc print data=count noobs;
run;
```

Explanation

This is an interesting application of the COMPRESS function. By computing the length of the string before and after removing the digits, this program sets the difference in the lengths to the number of digits in the original string. Notice the use of the LENGTHN function instead of the LENGTH function. When the COMPRESS function operates on the third observation (all digits), the result is a null string. The LENGTH function returns a value of 1 in this situation; the LENGTHN function returns a value of 0. See LENGTH and LENGTHN function descriptions for a detailed explanation.

```
Listing of Data Set COUNT

              Only_      Num_
String        letters    numerals

ABC123XYZ     ABCXYZ        3
XXXXX         XXXXX         0
12345                       5
1234X         X             4
```

Functions That Search for Characters

Functions in this category enable you to search a string for specific characters or for a character category (such as a digit). Some of these functions can also locate the first position in a string where a character does not meet a particular specification. Quite a few of the functions in this section were added with SAS®9, and they provide some new and useful capabilities.

The "ANY" Functions (ANYALNUM, ANYALPHA, ANYDIGIT, ANYPUNCT, ANYSPACE, ANYUPPER, and ANYLOWER)

This group of functions is described together because of the similarity of their use. These functions return the location of the first alphanumeric, letter, digit, etc. in a character value. Note that there are other "ANY" functions besides those presented here; these are the most common ones (for a complete list, see Product Documentation in the Knowledge Base, available at http://support.sas.com/documentation).

It is important to note that it might be necessary to use the TRIM function (or STRIP function) with the "ANY" and, especially, the "NOT" functions, since leading or trailing blanks might affect the results. For example, if X = "ABC   " (ABC followed by three blanks), Y = NOTALPHA(X) (return the location of the first character in X that is not a letter) will be 4, the location of the first trailing blank. Therefore, you might want to routinely use TRIM (or STRIP) like this:

```
Y = NOTALPHA(TRIM(X));
```

Note that there is a group of similar functions (NOTALPHA, NOTDIGIT, etc.) that work in a similar manner and are described together later in the next section. One program example follows the description of these seven functions.

Function: ANYALNUM

Purpose: To locate the first occurrence of an alphanumeric character (any uppercase or lowercase letter or number) and return its position. If none is found, the function returns a 0. With the use of an optional parameter, this function can begin searching at any position in the string and can also search from right to left, if desired.

Syntax: **ANYALNUM(*character-value <,start>*)**

character-value is any SAS character value.

start is an optional parameter that specifies the position in the string to begin the search. If it is omitted, the search starts at the beginning of the string. If it is non-zero, the search begins at the position in the string of the absolute value of the number (starting from the left-most position in the string). If the start value is positive, the search goes from left to right; if the value is negative, the search goes from right to left. A negative value larger than the length of the string results in a scan from right to left, starting at the end of the string. If the value of *start* is a positive number longer than the length of the string, or if it is 0, the function returns a 0.

Examples

For these examples, STRING = "ABC 123 ?xyz_n_".

Function	Returns
ANYALNUM(STRING)	1 (position of "A")
ANYALNUM("??$$%%")	0 (no alpha-numeric characters)
ANYALNUM(STRING,5)	5 (position of "1")
ANYALNUM(STRING,-4)	3 (position of "C")
ANYALNUM(STRING,6)	6 (position of "2")

Function: ANYALPHA

Purpose: To locate the first occurrence of an alpha character (any uppercase or lowercase letter) and return its position. If none is found, the function returns a 0. With the use of an optional parameter, this function can begin searching at any position in the string and can also search from right to left, if desired.

Syntax: ANYALPHA(*character-value* <,*start*>)

character-value is any SAS character value.

start is an optional parameter that specifies the position in the string to begin the search. If it is omitted, the search starts at the beginning of the string. If it is non-zero, the search begins at the position in the string of the absolute value of the number (starting from the left-most position in the string). If the start value is positive, the search goes from left to right; if the value is negative, the search goes from right to left. A negative value larger than the length of the string results in a scan from right to left, starting at the end of the string. If the value of *start* is a positive number longer than the length of the string, or if it is 0, the function returns a 0.

Examples

For these examples, STRING = "ABC 123 ?xyz_n_".

Function	Returns
ANYALPHA(STRING)	1 (position of "A")
ANYALPHA("??$$%%")	0 (no alpha characters)
ANYALPHA(STRING,5)	10 (position of "x")
ANYALPHA(STRING,-4)	3 (position of "C")
ANYALPHA(STRING,6)	10 (position of "x")

Function: ANYDIGIT

Purpose: To locate the first occurrence of a digit (numeral) and return its position. If none is found, the function returns a 0. With the use of an optional parameter, this function can begin searching at any position in the string and can also search from right to left, if desired.

Syntax: **ANYDIGIT(*character-value* <,*start*>)**

character-value is any SAS character value.

start is an optional parameter that specifies the position in the string to begin the search. If it is omitted, the search starts at the beginning of the string. If it is non-zero, the search begins at the position in the string of the absolute value of the number (starting from the left-most position in the string). If the start value is positive, the search goes from left to right; if the value is negative, the search goes from right to left. A negative value larger than the length of the string results in a scan from right to left, starting at the end of the string. If the value of start is a positive number longer than the length of the string, or if it is 0, the function returns a 0.

Examples

For these examples, STRING = "ABC 123 ?xyz_n_".

Function	Returns
ANYDIGIT(STRING)	5 (position of "1")
ANYDIGIT("??$$%%")	0 (no digits)
ANYDIGIT(STRING,5)	5 (position of "1")
ANYDIGIT(STRING,-4)	0 (no digits from position 4 to 1)
ANYDIGIT(STRING,6)	6 (position of "2")

Function: **ANYPUNCT**

Purpose: To locate the first occurrence of a punctuation character and return its position. If none is found, the function returns a 0. With the use of an optional parameter, this function can begin searching at any position in the string and can also search from right to left, if desired.

In the ASCII character set, the following characters are considered punctuation:

! " # $ % & ' ( ) * + , - . / : ;
< = > ? @ [\ ] ^ _ ` { | } ~

Syntax: **ANYPUNCT(*character-value* <,*start*>)**

character-value is any SAS character value.

start is an optional parameter that specifies the position in the string to begin the search. If it is omitted, the search starts at the beginning of the string. If it is non-zero, the search begins at the position in the string of the absolute value of the number (starting from the left-most position in the string). If the start value is positive, the search goes from left to right; if the value is negative, the search goes from right to left. A negative value larger than the length of the string results in a scan from right to left, starting at the end of the string. If the value of *start* is a positive number longer than the length of the string, or if it is 0, the function returns a 0.

Examples

For these examples, STRING = "A!C 123 ?xyz_n_".

Function	Returns
ANYPUNCT(STRING)	2 (position of "!")
ANYPUNCT("??$$%%")	1 (position of "?")
ANYPUNCT(STRING,5)	9 (position of "?")
ANYPUNCT(STRING,-4)	2 (starts at position 4 and goes left, position of "!")
ANYPUNCT(STRING,-3)	2 (starts at position 3 and goes left, position of "!")

Function: **ANYSPACE**

Purpose: To locate the first occurrence of a white space character (a blank, horizontal or vertical tab, carriage return, linefeed, and form-feed) and return its position. If none is found, the function returns a 0. With the use of an optional parameter, this function can begin searching at any position in the string and can also search from right to left, if desired.

Syntax: **ANYSPACE(*character-value <,start>*)**

character-value is any SAS character value.

start is an optional parameter that specifies the position in the string to begin the search. If it is omitted, the search starts at the beginning of the string. If it is non-zero, the search begins at the position in the string of the absolute value of the number (starting from the left-most position in the string). If the start value is positive, the search goes from left to right; if the value is negative, the search goes from right to left. A negative value larger than the length of the string results in a scan from right to left, starting at the end of the string. If the value of *start* is a positive number longer than the length of the string, or if it is 0, the function returns a 0.

Examples

For these examples, STRING = "ABC 123 ?xyz_n_".

Function	Returns
ANYSPACE(STRING)	4 (position of the first blank)
ANYSPACE("??$$%%")	0 (no spaces)
ANYSPACE(STRING,5)	8 (position of the second blank)
ANYSPACE(STRING,-4)	4 (position of the first blank)

Function: ANYUPPER

Purpose: To locate the first occurrence of an uppercase letter and return its position. If none is found, the function returns a 0. With the use of an optional parameter, this function can begin searching at any position in the string and can also search from right to left, if desired.

Syntax: ANYUPPER(*character-value* <,*start*>)

character-value is any SAS character value.

start is an optional parameter that specifies the position in the string to begin the search. If it is omitted, the search starts at the beginning of the string. If it is non-zero, the search begins at the position in the string of the absolute value of the number (starting from the left-most position in the string). If the start value is positive, the search goes from left to right; if the value is negative, the search goes from right to left. A negative value larger than the length of the string results in a scan from right to left, starting at the end of the string. If the value of *start* is a positive number longer than the length of the string, or if it is 0, the function returns a 0.

Examples

For these examples, STRING = "abcABC123".

Function	Returns
ANYUPPER(STRING)	4 (position of "A")
ANYUPPER("abc123")	0 (no uppercase characters)
ANYUPPER(STRING,5)	5 (position of "B")
ANYUPPER(STRING,-9)	6 (position of "C")

Function: ANYLOWER

Purpose: To locate the first occurrence of a lowercase letter and return its position. If none is found, the function returns a 0. With the use of an optional parameter, this function can begin searching at any position in the string and can also search from right to left, if desired.

Syntax: ANYLOWER(*character-value* <,*start*>)

character-value is any SAS character value.

start is an optional parameter that specifies the position in the string to begin the search. If it is omitted, the search starts at the beginning of the string. If it is non-zero, the search begins at the position in the string of the absolute value of the number (starting from the left-most position in the string). If the start value is positive, the search goes from left to right; if the value is negative, the search goes from right to left. A negative value larger than the length of the string results in a scan from right to left, starting at the end of the string. If the value of *start* is a positive number longer than the length of the string, or if it is 0, the function returns a 0.

Examples

For these examples, STRING = "abcABC123".

Function	Returns
ANYLOWER(STRING)	1 (position of "a")
ANYLOWER("ABC123")	0 (no lowercase characters)
ANYLOWER(STRING,3)	3 (position of "c")
ANYLOWER(STRING,-9)	3 (position of "c")

Program 1.11: Demonstrating the "ANY" character functions

```
***Primary functions: ANYALNUM, ANYALPHA, ANYDIGIT, ANYPUNCT, ANYSPACE
                       ANYUPPER, and ANYLOWER;

data anywhere;
   input String $char20.;
   Alpha_num   = anyalnum(String);
   Alpha_num_9 = anyalnum(String,-999);
   Alpha       = anyalpha(String);
   Alpha_5     = anyalpha(String,-5);
   Digit       = anydigit(String);
   Digit_9     = anydigit(String,-999);
   Punct       = anypunct(String);
   Space       = anyspace(String);
   Up          = anyupper(String);
   Low         = anylower(String);
datalines;
Once upon a time 123
HELP!
987654321
UPPER and lower
;
title "Listing of Data Set ANYWHERE";
proc print data=anywhere noobs heading=h;
run;
```

Explanation

Each of these "ANY" functions works in a similar manner, the only difference being in the
types of character values it is searching for. The two statements using a starting value of
–999 demonstrate an easy way to search from right to left, without having to know the length
of the string (assuming that you don't have any strings longer than 999, in which case
you could choose a larger number). Functions such as ANYALPHA and ANYDIGIT can be
very useful for extracting values from strings where the positions of digits or letters are not
fixed. An alternative to using this group of functions would be to use Perl regular
expressions. See the following chapter for a complete discussion of regular expressions.
Notice in the following listing that the positions of the first space in lines two and three are 6
and 10, respectively. These are the positions of the first trailing blank in each of the two
strings (remember that the length of STRING is 20).

```
Listing of Data Set ANYWHERE

                    Alpha_    Alpha_
String                num     num_9    Alpha

Once upon a time 123     1       20       1
HELP!                    1        4       1
987654321                1        9       0
UPPER and lower          1       15       1



Alpha_5    Digit    Digit_9    Punct    Space    Up    Low

   4         18        20        0        5      1      2
   4          0         0        5        6      1      0
   0          1         9        0       10      0      0
   5          0         0        0        6      1      7
```

Program 1.12: Using the functions ANYDIGIT and ANYSPACE to find the first number in a string

```
***Primary functions: ANYDIGIT and ANYSPACE
***Other functions: INPUT and SUBSTR;

data search_num;
    input String $60.;
    Start = anydigit(String);
    End = anyspace(String,Start);
    if Start then
        Num = input(substr(String,Start,End-Start),9.);
datalines;
This line has a 56 in it
two numbers 123 and 456 in this line
No digits here
;
title "Listing of Data Set SEARCH_NUM";
proc print data=search_num noobs;
run;
```

Explanation

This program identifies the first number in any line of data that contains a numeric value (followed by one or more blanks). The ANYDIGIT function determines the position of the first digit of the number; the ANYSPACE function searches for the first blank following the number (the starting position of this search is the position of the first digit). The SUBSTR function extracts the digits (starting at the value of START with a length determined by the difference between END and START). Finally, the INPUT function performs the character-to-numeric conversion. You can also write the statement IF START as IF START GT 0 if you want. However, any numeric value that is not zero or missing is considered "true." The following listing shows that this program works as expected.

```
Listing of Data Set SEARCH_NUM

String                                 Start   End    Num

This line has a 56 in it                 17     19     56
two numbers 123 and 456 in this line     13     16     123
No digits here                            0      0      .
```

The "NOT" Functions (NOTALNUM, NOTALPHA, NOTDIGIT, NOTUPPER and NOTLOWER)

This group of functions is similar to the "ANY" functions (such as ANYALNUM, ANYALPHA, etc.) except that the functions return the position of the first character value that is **not** a particular value (alphanumeric, character, digit, or uppercase character). Note that this is not a complete list of the "NOT" functions. For a complete list, see Product Documentation in the Knowledge Base, available at http://support.sas.com/documentation.

As with the "ANY" functions, there is an optional parameter that specifies where to start the search and in which direction to search.

Function: NOTALNUM

Purpose: To determine the position of the first character in a string that is not an alphanumeric (any uppercase or lowercase letter or a number). If none is found, the function returns a 0. With the use of an optional parameter, this function can begin searching at any position in the string and can also search from right to left, if desired.

Syntax: NOTALNUM(*character-value <,start>*)

character-value is any SAS character value.

start is an optional parameter that specifies the position in the string to begin the search. If it is omitted, the search starts at the beginning of the string. If it is non-zero, the search begins at the position in the string of the absolute value of the number (starting from the left-most position in the string). If the start value is positive, the search goes from left to right; if the value is negative, the search goes from right to left. A negative value larger than the length of the string results in a scan from right to left, starting at the end of the string. If the value of `start` is a positive number longer than the length of the string, or if it is 0, the function returns a 0.

Examples

For these examples, STRING = "ABC 123 ?xyz_n_".

Function	Returns
NOTALNUM(STRING)	4 (position of the 1st blank)
NOTALNUM("Testing123")	0 (all alpha-numeric values)
NOTALNUM("??$$%%")	1 (position of the "?")
NOTALNUM(STRING,5)	8 (position of the 2nd blank)
NOTALNUM(STRING,-6)	4 (position of the 1st blank)
NOTALNUM(STRING,8)	9 (position of the "?")

Function: **NOTALPHA**

Purpose: To determine the position of the first character in a string that is not an uppercase or lowercase letter (alpha character). If none is found, the function returns a 0. With the use of an optional parameter, this function can begin searching at any position in the string and can also search from right to left, if desired.

Syntax: NOTALPHA(*character-value <,start>*)

character-value is any SAS character value.

start is an optional parameter that specifies the position in the string to begin the search. If it is omitted, the search starts at the beginning of the string. If it is non-zero, the search begins at the position in the string of the absolute value of the number (starting from the left-most position in the string). If the start value is positive, the search goes from left to right; if the value is negative, the search goes from right to left. A negative value larger than the length of the string results in a scan from right to left, starting at the end of the string. If the value of *start* is a positive number longer than the length of the string, or if it is 0, the function returns a 0.

Examples

For these examples, STRING = "ABC 123 ?xyz_n_".

Function	Returns
NOTALPHA (STRING)	4 (position of 1st blank)
NOTALPHA ("ABCabc")	0 (all alpha characters)
NOTALPHA ("??$$%%")	1 (position of first "?")
NOTALPHA (STRING,5)	5 (position of "1")
NOTALPHA (STRING,-10)	9 (start at position 10 and search left, position of "?")
NOTALPHA (STRING,2)	4 (position of 1st blank)

Function: **NOTDIGIT**

Purpose: To determine the position of the first character in a string that is not a digit. If none is found, the function returns a 0. With the use of an optional parameter, this function can begin searching at any position in the string and can also search from right to left, if desired.

Syntax: NOTDIGIT(*character-value* <,*start*>)

character-value is any SAS character value.

start is an optional parameter that specifies the position in the string to begin the search. If it is omitted, the search starts at the beginning of the string. If it is non-zero, the search begins at the position in the string of the absolute value of the number (starting from the left-most position in the string). If the start value is positive, the search goes from left to right; if the value is negative, the search goes from right to left. A negative value larger than the length of the string results in a scan from right to left, starting at the end of the string. If the value of *start* is a positive number longer than the length of the string, or if it is 0, the function returns a 0.

Examples

For these examples, STRING = "ABC 123 ?xyz_n_".

Function	Returns
NOTDIGIT(STRING)	1 (position of "A")
NOTDIGIT("123456")	0 (all digits)
NOTDIGIT("??$$%%")	1 (position of "?")
NOTDIGIT(STRING,5)	8 (position of 2nd blank)
NOTDIGIT(STRING,-6)	4 (position of 1st blank)
NOTDIGIT(STRING,6)	8 (position of 2nd blank)

Function: NOTUPPER

Purpose: To determine the position of the first character in a string that is not an uppercase letter. If none is found, the function returns a 0. With the use of an optional parameter, this function can begin searching at any position in the string and can also search from right to left, if desired.

Syntax: NOTUPPER(*character-value* <,*start*>)

character-value is any SAS character value.

start is an optional parameter that specifies the position in the string to begin the search. If it is omitted, the search starts at the beginning of the string. If it is non-zero, the search begins at the position in the string of the absolute value of the number (starting from the left-most position in the string). If the start value is positive, the search goes from left to right; if the value is negative, the search goes from right to left. A negative value larger than the length of the string results in a scan from right to left, starting at the end of the string. If the value of *start* is a positive number longer than the length of the string, or if it is 0, the function returns a 0.

Examples

For these examples, STRING = "ABC 123 ?xyz_n_".

Function	Returns
NOTUPPER("ABCDabcd")	5 (position of "a")
NOTUPPER("ABCDEFG")	0 (all uppercase characters)
NOTUPPER(STRING)	4 (position of 1st blank)
NOTUPPER("??$$%%")	1 (position of "?")
NOTUPPER(STRING,5)	5 (position of "1")
NOTUPPER(STRING,-6)	6 (position of "2")
NOTUPPER(STRING,6)	6 (position of "2")

Function: NOTLOWER

Purpose: To determine the position of the first character in a string that is not a lowercase letter. If none is found, the function returns a 0. With the use of an optional parameter, this function can begin searching at any position in the string and can also search from right to left, if desired.

Syntax: NOTLOWER(*character-value* <,*start*>)

character-value is any SAS character value.

start is an optional parameter that specifies the position in the string to begin the search. If it is omitted, the search starts at the beginning of the string. If it is non-zero, the search begins at the position in the string of the absolute value of the number (starting from the left-most position in the string). If the start value is positive, the search goes from left to right; if the value is negative, the search goes from right to left. A negative value larger than the length of the string results in a scan from right to left, starting at the end of the string. If the value of *start* is a positive number longer than the length of the string, or if it is 0, the function returns a 0.

Examples

For these examples, STRING = "abc 123 ?XYZ_n_".

Function	Returns
NOTLOWER("ABCDabcd")	1 (position of "A")
NOTLOWER("abcdefg")	0 (all lowercase characters)
NOTLOWER(STRING)	4 (position of blank)
NOTLOWER("??$$%%")	1 (position of "?")
NOTLOWER(STRING,5)	5 (position of "1")
NOTLOWER(STRING,-6)	6 (position of "2")
NOTLOWER(STRING,6)	6 (position of "2")

Program 1.13: Demonstrating the "NOT" character functions

```
***Primary functions: NOTALNUM, NOTALPHA, NOTDIGIT, NOTUPPER, and
NOTLOWER;

data negative;
   input String $5.;
   Not_alpha_numeric = notalnum(String);
   Not_alpha         = notalpha(String);
   Not_digit         = notdigit(String);
   Not_upper         = notupper(String);
   Not_lower         = notlower(String);
datalines;
ABCDE
abcde
abcDE
12345
:#$%&
ABC
;
title "Listing of Data Set NEGATIVE";
proc print data=negative noobs;
run;
```

Explanation

This straightforward program demonstrates each of the "NOT" character functions. As with most character functions, be careful with trailing blanks. Notice that the last observation ("ABC") contains only three characters, but since STRING is read with a $5. informat, there are two trailing blanks following the letters "ABC". That is the reason you obtain a value of 4 for all the functions except NOTDIGIT and NOTLOWER (the first character is not a digit or a lowercase character). A listing of the data set NEGATIVE is shown next:

```
Listing of Data Set NEGATIVE

             Not_alpha_     Not_     Not_     Not_     Not_
String         numeric     alpha    digit    upper    lower

ABCDE            0           0        1        0        1
abcde            0           0        1        1        0
abcDE            0           0        1        1        4
12345            0           1        0        1        1
:#$%&            1           1        1        1        1
ABC              4           4        1        4        1
```

FIND, FINDC, and FINDW

This group of functions shares some similarities to the INDEX, INDEXC, and INDEXW functions. FIND and INDEX both search a string for a given substring. FINDC and INDEXC both search for individual characters. FINDW and INDEXW both search for "words" (a string bounded by delimiters). However, all of the FIND functions have some additional capability over their counterparts. For example, the set of FIND functions has the ability to declare a starting position for the search, set the direction of the search, and ignore case or trailing blanks.

Function: FIND

Purpose: To locate a substring within a string. With optional arguments, you can define the starting point for the search, set the direction of the search, and ignore case or trailing blanks.

Syntax: FIND(*character-value, find-string <,'modifiers'> <,start>*)

character-value is any SAS character value.

find-string is a character variable or string literal that contains one or more characters that you want to search for. The function returns the first position in the *character-value* that contains the *find-string*. If the *find-string* is not found, the function returns a 0.

The following *modifiers* (in uppercase or lowercase), placed in single or double quotation marks, can be used with FIND:

i ignore case.

t ignore trailing blanks in both the character variable and the *find-string*.

start is an optional parameter that specifies the position in the string to begin the search. If it is omitted, the search starts at the beginning of the string. If it is non-zero, the search begins at the position in the string of the absolute value of the number (starting from the left-most position in the string). If the start value is positive, the search goes from left to right; if the value is negative, the search goes from right to left. A negative value larger than the length of the string results in a scan from right to left, starting at the end of the string. If the value of *start* is a positive number longer than the length of the string, or if it is 0, the function returns a 0.

Note: You can switch the positions of *start* and *modifiers* and the function will work the same. You can also use either one of these arguments without the other (as the third argument to the function).

Examples

For these examples, STRING1 = "Hello hello goodbye" and STRING2 = "hello".

Function	Returns
FIND(STRING1, STRING2)	7 (position of "h" in hello)
FIND(STRING1, STRING2, 'i')	1 (position of "H" in Hello)
FIND(STRING1,"bye")	17 (position of "b" in goodbye)
FIND("abcxyzabc","abc",4)	7 (position of second "a")
FIND(STRING1, STRING2, 'i', -99)	7 (position of "h" in hello)

Function: **FINDC**

Purpose: To locate a character that appears or does not appear within a string. With optional arguments, you can define the starting point for the search, set the direction of the search, ignore case or trailing blanks, or look for characters except the ones listed.

Syntax: FINDC(*character-value, find-characters*
      <,'modifiers'> <,start>)

character-value is any SAS character value.

find-characters is a list of one or more characters that you want to search for.

The function returns the first position in the *character-value* that contains one of the *find-characters*. If none of the characters are found, the function returns a 0. With an optional argument, you can have the function return the position in a character string of a character that is not in the *find-characters* list.

The following *modifiers* (in uppercase or lowercase), placed in single or double quotation marks, can be used with FINDC:

   i ignore case.

   t ignore trailing blanks in both the character variable and the *find-characters*.

   k count only characters that are not in the list of *find-characters*.

   o process the *modifiers* and *find-characters* only once for a specific call to the function. In subsequent calls, changes to these arguments will have no effect. This might improve performance if the *find-characters* are constants or literals.

   a add uppercase and lowercase letters to the list of characters to find.

   d add digits to the list of characters to find.

start is an optional parameter that specifies the position in the string to begin the search. If it is omitted, the search starts at the beginning of the string. If it is non-zero, the search begins at the position in the string of the absolute value of the number (starting from the left-most position in the string). If the start value is positive, the search goes from left to right; if the value is negative, the search goes from right to left. A negative value larger than the length of the string results in a scan from right to left, starting at the end of the string. If the value of *start* is a positive number longer than the length of the string, or if it is 0, the function returns a 0.

Note: You can switch the positions of *start* and *modifiers* and the function will work the same. You can also use either one of these arguments without the other (as the third argument to the function).

Examples

For these examples, STRING1 = "Apples and Books" and STRING2 = "abcde".

Function	Returns
FINDC(STRING1, STRING2)	5 (position of "e" in Apples)
FINDC(STRING1, STRING2, 'i')	1 (position of "A" in Apples)
FINDC(STRING1,"aple",'ki')	1 (position of "A" in Apple)
FINDC("abcxyzabc","abc",4)	7 (position of second "a")

Program 1.14: Using the FIND and FINDC functions to search for strings and characters

```
***Primary functions: FIND and FINDC;

data find_vowel;
   input @1 String $20.;
   Pear = find(String,'pear');
   Pos_vowel = findc(String,'aeiou','i');
   Upper_vowel = findc(String,'AEIOU');
   Not_vowel = findc(String,'aeiou','ik');
datalines;
XYZABCabc
XYZ
Apple and Pear
;
```

```
title "Listing of Data Set FIND_VOWEL";
proc print data=find_vowel noobs;
run;
```

Explanation

The FIND function returns the position of the characters "Pear" in the variable STRING. Since the i modifier is not used, the search is case sensitive. The first use of the FINDC function looks for any uppercase or lowercase vowel in the string (because of the i modifier). The next statement, without the i modifier, locates only lowercase vowels. Finally, the k modifier in the last FINDC function reverses the search to look for the first character that is not a vowel (uppercase or lowercase because of the i modifier).

Here is the output:

```
Listing of Data Set FIND_VOWEL


                             Pos_      Upper_     Not_
String            Pear      vowel      vowel      vowel

XYZABCabc          0          4          4          1
XYZ                0          0          0          1
Apple and Pear     0          1          1          2
```

Here is another example where the FINDC functions makes child's play out of what might seem, at first glance, to be a hard problem.

Program 1.15: Converting numeric values of mixed units (e.g., kg and lbs) to a single numeric quantity

```
***Primary functions: COMPRESS, FINDC, INPUT
***Other function: ROUND;

data heavy;
   input Char_wt $ @@;
   Weight = input(compress(Char_wt,,'kd'),8.);
   if findc(Char_wt,'k','i') then Weight = 2.22 * Weight;
   Weight = round(Weight);
datalines;
60KG 155 82KG 54kg 98
;
title "Listing of Data Set HEAVY";
proc print data=heavy noobs;
   var Char_wt Weight;
run;
```

Explanation

The data lines contain numbers in kilograms, followed by the abbreviation KG, or in pounds (no units used). As with most problems of this type, when you are reading a combination of numbers and characters, you usually need to first read the value as a character. Here the COMPRESS function is used to extract the digits from the WEIGHT value. The INPUT function does its usual job of character-to-numeric conversion. If the FINDC function returns any value other than a 0, the letter K (uppercase or lowercase) was found in the string and the WEIGHT value is converted from KG to pounds. Finally, the value is rounded to the nearest pound, using the ROUND function. The listing of data set HEAVY follows:

```
Listing of Data Set HEAVY

Char_wt     Weight

 60KG          133
 155           155
 82KG          182
 54kg          120
 98             98
```

Program 1.16: Searching for one of several characters in a character variable

```
***Primary function: FINDC;

data check;
   input Tag_number $ @@;
   ***if the Tag number contains an x, y, or z, it indicates
      an international destination, otherwise, the destination
      is domestic;
   if findc(tag_number,'xyz','i') then
      Destination = 'International';
   else Destination = 'Domestic';
datalines;
T123 ty333 1357Z UZYX 888 ABC
;
title "Listing of Data Set CHECK";
proc print data=check noobs;
   id Tag_number;
   var Destination;
run;
```

Explanation

If an X, Y, or Z (uppercase or lowercase) is found in the variable TAG_NUMBER, the function returns a number greater than 0 (true) and DESTINATION will be set to INTERNATIONAL. As you can see in the following listing, this use of the FINDC function works as advertised.

```
Listing of Data Set CHECK

 Tag_
number     Destination

T123       Domestic
ty333      International
1357Z      International
UZYX       International
888        Domestic
ABC        Domestic
```

Program 1.17: Demonstrating the o modifier with FINDC

```
***Primary function: FINDC;

data o_modifier;
   input String       $15.
         @16 Look_for $1.;
   Position = findc(String,Look_for,'io');
datalines;
Capital A here A
Lower a here   X
Apple          B
;
title "Listing of Data Set O_MODIFIER";
proc print data=o_modifier noobs heading=h;
run;
```

Explanation

The o modifier is usually used to slightly improve efficiency of a program where the find character(s) is expressed as a constant or literal (i.e., do not change) as opposed to a variable.

This program demonstrates what happens if you try to change the value of the *find-string* and use the o modifier. In the first iteration of the DATA step, the value of LOOK_FOR is an uppercase or lowercase A. Since the o modifier was used, changing the value of LOOK_FOR in the next two observations has no effect; the function continues to look for the letter A (uppercase or lowercase).

Note that another use of FINDC in this DATA step would not be affected by the previous use of the o modifier, even if the name of the variable (in this case POSITION) were the same. The o modifier is most likely useful in reducing processing time when looping through multiple strings, looking for the same string with the same modifiers. The following listing of data set O_MODIFIER shows that, even though the LOOK_FOR value was changed to X in the second observation and B in the third observation, the function continues to search for the letter A.

```
Listing of Data Set O_MODIFIER

String            Look_for    Position

Capital A here       A            2
Lower a here         X            7
Apple                B            1
```

Function: **FINDW**

Purpose: To search a string for a word, defined as a group of letters separated on both ends by a word boundary (a space, the beginning of a string, or the end of the string). Note that punctuation is not considered a word boundary.

Syntax: FINDW(*char-value, word <, delim>*) or
FINDW(*char_value, word, delim, modifier<,start>*)

char-value is any SAS character value.

word is the word for which you want to search.

delim specifies values to be used as delimiters. The default delimiters in ASCII are blank, ! $ % & ( ) * + , - . / ; < ^ and |. In EBCDIC, they are blank, ! $ % & ( ) * + , - . / ; < ¬ | and ¢.

modifier is an optional argument that changes the search conditions (the most popular modifier is i (ignore case). If you supply a modifier, you must also supply a delimiter. (Since modifiers and delimiters are both character values, the function would assume you were supplying a delimiter and not a modifier if you did not specify a value for the delimiter.)

start gives a starting position for the search. If negative, search goes from right to left, starting at the position of the absolute value of *start*.

The function returns the first position in the *char-value* that contains the *word*. If the *word* is not found, the function returns a 0. The FINDW function can be complicated, especially when using some of the modifiers or a negative starting position. We recommend that you see Product Documentation in the Knowledge Base, available at http://support.sas.com/documentation, for details.

Examples

For these examples, STRING1 = "Hello hello goodbye" and STRING2 = "hello".

Function	Result
FINDW(STRING1,STRING2)	7 (the word "hello")
FINDW(STRING1,STRING2,' ','i')	1 (the word "Hello")
FINDW(STRING1,"good")	0 (no word boundary)
FINDW("1 and 2 and 3","and",7)	9 (2nd "and")
FINDW("one:two:three","two",':')	5 (":" defined as a delimiter)

Program 1.18: Searching for a word using the FINDW function

```
***Primary functions: FIND and FINDW;

data find_word;
   input String $40.;
   Position_w = findw(String,"the");
   Position   = find(String,"the");
datalines;
there is a the in this line
ends in the
ends in the.
none here
;
title "Listing of Data Set FIND_WORD";
proc print data=find_word;
run;
```

Explanation

This program demonstrates the difference between FIND and FINDW. Notice in the first observation in the following listing, the FIND function returns a 1 because the letters "the" as part of the word "there" begin the string. Since the FINDW function needs white space at the beginning and end of a string to delimit a word, it returns a 12, the position of the word "the" in the string. Observation 3 emphasizes the fact that a punctuation mark does serve as a word separator. Finally, since the string "the" does not appear anywhere in the fourth observation, both functions return a 0. Here is the listing:

```
Listing of Data Set FIND_WORD

                                   Position_
Obs    String                         W         Position

1      there is a the in this line    12            1
2      ends in the                     9            9
3      ends in the.                    9            9
4      none here                       0            0
```

INDEX, INDEXC, and INDEXW

This group of functions all search a string for a substring of one or more characters. INDEX and INDEXW are similar, the difference being that INDEXW looks for a word (defined as a string bounded by spaces or the beginning or end of the string), while INDEX simply searches for the designated substring. INDEXC searches for one or more individual characters and always searches from left to right. Note that these three functions are all case sensitive.

All three of these functions have now been improved and replaced by FIND, FINDC, and FINDW. The three INDEX functions are included in this book since they were widely used in older programs. We highly recommend that in any new program, you use the group of FIND functions instead of the INDEX functions. The FIND functions can do anything the older INDEX functions did, plus they have the advantage of allowing you to specify a starting position and, especially important, use modifiers (my favorite is the i modifier to ignore case).

Function: INDEX

Purpose: To locate the starting position of a substring in a string.

Syntax: INDEX(*character-value, find-string*)

character-value is any SAS character value.

find-string is a character variable or string literal that contains the substring for which you want to search.

The function returns the first position in the *character-value* that contains the *find-string*. If the *find-string* is not found, the function returns a 0.

Examples

For these examples, STRING = "ABCDEFG".

Function	Returns
INDEX(STRING,'C')	3 (the position of the "C")
INDEX(STRING,'DEF')	4 (the position of the "D")
INDEX(STRING,'X')	0 (no "X" in the string)
INDEX(STRING,'ACE')	0 (no "ACE" in the string)

Function: INDEXC

Purpose: To search a character string for one or more characters. The INDEXC function works in a similar manner to the INDEX function, with the difference being it can be used to search for any one in a list of character values.

Syntax: INDEXC(*character-value, 'char1','char2','char3', ...)*

or

INDEXC(*character-value*, *'char1char2char3. . .'*)

character-value is any SAS character value.

char1, char2, . . . are individual character values that you want to search for in the *character-value*.

The INDEXC function returns the first occurrence of any of the char1, char2, etc. values in the string. If none of the characters is found, the function returns a 0.

Examples

For these examples, STRING = "ABCDEFG".

Function	Returns
INDEXC(STRING,'F','C','G')	3 (position of the "C")
INDEXC(STRING, 'FCG')	3 (position of the "C")
INDEXC(STRING,'X','Y','Z')	0 (no "X", "Y", or "Z" in STRING)

Note: It makes no difference if you list the search characters as 'FCG' or 'F','C','G'.

Program 1.19: Reading dates in a mixture of formats

```
***Primary function: INDEXC
***Other function: INPUT;

***Note: SAS 9 has some enhanced date reading ability;

***Program to read mixed dates;
data mixed_dates;
   input @1 Dummy $15.;
   if indexc(dummy,'/-:') then Date = input(Dummy,mmddyy10.);
   else Date = input(Dummy,date9.);
   format Date worddate.;
datalines;
10/21/1946
06JUN2002
5-10-1950
7:9:57
;
title "Listing of Data Set MIXED_DATES";
proc print data=mixed_dates noobs;
run;
```

Explanation

In this somewhat trumped-up example, dates are entered either in mm/dd/yyyy or ddMONyyyy form. Also, besides a slash, dashes and colons are used. Any string that includes either a slash, dash, or colon is a date that needs the MMDDYY10. informat. Otherwise, the DATE9. informat is used.

The alert reader will realize that the ANYDTDTE10. informat (SAS®9) could read this raw data directly and create a SAS date value. Here is a listing of the data set MIXED_DATES:

```
Listing of Data Set MIXED_DATES

Dummy            Date

10/21/1946    October 21, 1946
06JUN2002        June 6, 2002
5-10-1950        May 10, 1950
7:9:57           July 9, 1957
```

Function: VERIFY

Purpose: To check whether a string contains any unwanted values.

Syntax: VERIFY(*character-value, verify-string*)

character-value is any SAS character value.

verify-string is a SAS character variable or a list of character values in quotation marks.

This function returns the first position in the *character-value* that is **not** present in the *verify-string*. If the *character-value* does not contain any characters other than those in the *verify-string*, the function returns a 0. Be especially careful to think about trailing blanks when using this function. If you have an 8-byte character variable equal to 'ABC' (followed by five blanks), and if the verify string is equal to 'ABC', the VERIFY function returns a 4, the position of the first blank (which is not present in the verify string). Therefore, you may need to use the TRIM function on either the character-value, the verify-string, or both.

Examples

For these examples, STRING = "ABCXABD" and V = "ABCDE".

Function	Returns
VERIFY(STRING,V)	4 ("X" is not in the verify string)
VERIFY(STRING,"ABCDEXYZ")	0 (no "bad" characters in STRING)
VERIFY(STRING,"ACD")	2 (position of the "B")
VERIFY("ABC ","ABC")	4 (position of the 1st blank)
VERIFY(TRIM("ABC "),"ABC")	0 (no invalid characters)

Program 1.20: Using the VERIFY function to check for invalid character data values

```
***Primary function: VERIFY;

data very_fi;
   input ID      $ 1-3
         Answer $ 5-9;
   P = verify(Answer,'ABCDE');
   OK = P eq 0;
datalines;
001 ACBED
002 ABXDE
003 12CCE
004 ABC E
;
title "listing of Data Set VERY_FI";
proc print data=very_fi noobs;
run;
```

Explanation

In this example, the only valid values for ANSWER are the uppercase letters A–E. Any time there are one or more invalid values, the result of the VERIFY function (variable P) will be a number from 1 to 5. The SAS statement that computes the value of the variable OK needs a word of explanation. First, the logical comparison P EQ 0 returns a value of true or false, which is equivalent to a 1 or 0. This value is then assigned to the variable OK. Thus, the variable OK is set to 1 for all valid values of ANSWER and to 0 for any invalid values. This use of the VERIFY function is very handy in some data cleaning applications. Here is a listing of data set VERI_FI:

```
listing of Data Set VERY_FI

ID     Answer    P    OK

001    ACBED     0    1
002    ABXDE     3    0
003    12CCE     1    0
004    ABC E     4    0
```

Functions That Extract Parts of Strings

The functions described in this section can extract parts of strings. When used on the left-hand side of the equal sign, the SUBSTR function can also be used to insert characters into specific positions of an existing string.

Function: SUBSTR

Purpose: To extract part of a string. When the SUBSTR function is used on the left side of the equal sign, it can place specified characters into an existing string.

Syntax: SUBSTR(*character-value, start <,length>*)

character-value is any SAS character value.

start is the starting position within the string.

length, if specified, is the number of characters to include in the substring. If this argument is omitted, the SUBSTR function will return all the characters from the start position to the end of the string.

If a length has not been previously assigned, the length of the resulting variable will be the length of the *character-value*.

Examples

For these examples, STRING = "ABC123XYZ".

Function	Returns
SUBSTR(STRING,4,2)	"12"
SUBSTR(STRING,4)	"123XYZ"
SUBSTR(STRING,LENGTH(STRING))	"Z" (last character in the string)

Program 1.21: Extracting portions of a character value and creating a character variable and a numeric value

```
***Primary function: SUBSTR
***Other function: INPUT;

data substring;
   input ID $ 1-9;
   length State $ 2;
   State = ID;
   Num = input(substr(ID,7,3),3.);
datalines;
NYXXXX123
NJ1234567
;
title 'Listing of Data Set SUBSTRING';
proc print data=substring noobs;
run;
```

Explanation

In this example, the ID contains both state and number information. The first two characters of the ID variable contain the state abbreviations and the last three characters represent digits that you want to use to create a numeric variable. A useful and efficient trick is used to obtain the state code. Since the length statement assigns a length of two for STATE, setting STATE equal to ID results in the first two characters in ID (the state codes) as the values for STATE. To obtain a numeric value from the last 3 bytes of the ID variable, it is necessary to first use the SUBSTR function to extract the three characters of interest and to then use the INPUT function to do the character to numeric conversion. A listing of data set SUBSTRING is shown next:

```
Listing of Data Set SUBSTRING

   ID        State   Num

NYXXXX123     NY     123
NJ1234567     NJ     567
```

Program 1.22: Extracting the last two characters from a string, regardless of the length

```
***Primary functions: LENGTH, SUBSTR;

data extract;
   input @1 String $20.;
        Last_two = substr(String,length(String)-1,2);
datalines;
ABCDE
AX12345NY
126789
;
title "Listing of Data Set EXTRACT";
proc print data=extract noobs;
run;
```

Explanation

This program demonstrates how you can use the LENGTH and SUBSTR functions together to extract portions of a string when the strings are of different or unknown lengths. To see how this program works, take a look at the first line of data. The LENGTH function will return a 5, and (5–1) = 4, the position of the next to the last (penultimate) character in STRING. See the following listing:

```
Listing of Data Set EXTRACT

String      Last_two

ABCDE          DE
AX12345NY      NY
126789         89
```

Program 1.23: Using the SUBSTR function to "unpack" a string

```
***Primary function: SUBSTR
***Other functions: INPUT;

data pack;
   input String $ 1-5;
datalines;
12345
8 642
;
data unpack;
   set pack;
   array x[5];
   do j = 1 to 5;
      x[j] = input(substr(String,j,1),1.);
   end;
   drop j;
run;
title "Listing of Data Set UNPACK";
proc print data=unpack noobs;
run;
```

Explanation

There are times when you want to store a group of one-digit numbers in a compact, space-saving way. In this example, you want to store five one-digit numbers. If you stored each one as an 8-byte numeric, you would need 40 bytes of storage for each observation. By storing the five numbers as a 5-byte character string, you need only 5 bytes of storage. However, you need to use CPU time to turn the character string back into the five numbers.

The key here is to use the SUBSTR function with the starting value as the index of a DO loop. As you pick off each of the numerals, you can use the INPUT function to do the character-to-numeric conversion. Notice that the ARRAY statement in this program does not include a list of variables. When this list is omitted and the number of elements is placed in parentheses, SAS automatically uses the array name followed by the numbers from 1 to n, where n is the number in parentheses. A listing of data set UNPACK follows:

```
Listing of Data Set UNPACK

String   x1    x2    x3    x4    x5

12345     1     2     3     4     5
8 642     8     .     6     4     2
```

Function: SUBSTR (on the left-hand side of the equal sign)

As we mentioned in the description of the SUBSTR function, there is an interesting and useful way it can be used—on the left-hand side of the equal sign.

Purpose: To place one or more characters into an existing string.

Syntax: SUBSTR(*character-value, start <, length>*) = *character-value*

character-value is any SAS character value.

start is the starting position in a string where you want to place the new characters.

length is the number of characters to be placed in that string. If *length* is omitted, all the characters on the right-hand side of the equal sign replace the characters in *character-value*.

Examples

In these examples, EXISTING = "ABCDEFGH", NEW = "XY".

Function	Returns
SUBSTR(EXISTING,3,2) = NEW	EXISTING is now = "ABXYEFGH"
SUBSTR(EXISTING,3,1) = "*"	EXISTING is now = "AB*DEFGH"

Program 1.24: Demonstrating the SUBSTR function on the left-hand side of the equal sign

```
***Primary function: SUBSTR
***Other function: PUT;

data stars;
   input SBP DBP @@;
   length SBP_chk DBP_chk $ 4;
   SBP_chk = put(SBP,3.);
   DBP_chk = put(DBP,3.);
   if SBP gt 160 then substr(SBP_chk,4,1) = '*';
   if DBP gt 90 then substr(DBP_chk,4,1) = '*';
datalines;
120 80 180 92 200 110
;
title "Listing of data set STARS";
proc print data=stars noobs;
run;
```

Explanation

In this program, you want to flag high values of systolic and diastolic blood pressure by placing an asterisk after the value. Notice that the variables SBP_CHK and DBP_CHK are both assigned a length of 4 by the LENGTH statement. The fourth position needs to be there in case you want to place an asterisk in that position, to flag the value as abnormal. The PUT function places the numerals of the blood pressures into the first 3 bytes of the corresponding character variables. Then, if the value is above the specified level, an asterisk is placed in the fourth position of these variables.

```
Listing of data set STARS

SBP    DBP    SBP_chk    DBP_chk

120     80     120         80
180     92     180*        92*
200    110     200*       110*
```

Function: **SUBSTRN**

Purpose: This function serves the same purpose as the SUBSTR function with a few added features. Unlike the SUBSTR function, the starting position and the length arguments of the SUBSTRN function can be 0 or negative without causing an error. In particular, if the length is 0, the function returns a string of 0 length. This is particularly useful when you are using regular expression functions where the length parameter may be 0 when a pattern is not found (with the PRXSUBSTR function, for example). You can use the SUBSTRN function in any application where you would use the SUBSTR function. The effect of 0 or negative parameters is discussed in the following description of the arguments.

Syntax: SUBSTRN(*character-value, start <, length>*)

character-value is any SAS character variable.

start is the starting position in the string. If this value is non-positive, the function returns a substring starting from the first character in *character-value* (the length of the substring will be computed by counting, starting from the value of *start*). See the following program.

length is the number of characters in the substring. If this value is non-positive (in particular, 0), the function returns a string of length 0. If this argument is omitted, the SUBSTRN function will return all the characters from the start position to the end of the string.

If a length has not been previously assigned, the length of the resulting variable will be the length of the *character-value*.

Examples

For these examples, STRING = "ABCDE".

Function	Returns
SUBSTRN(STRING,2,3)	"BCD"
SUBSTRN(STRING,-1,4)	"AB"
SUBSTRN(STRING,4,5)	"DE"
SUBSTRN(STRING,3,0)	string of zero length

Program 1.25: Demonstrating the unique features of the **SUBSTRN** function

```
***Primary function: SUBSTRN;

title "Demonstrating the SUBSTRN Function";
data hoagie;
   String = 'abcdefghij';
   length Result $5.;
   Result = substrn(String,2,5);
   Sub1 = substrn(String,-1,4);
   Sub2 = substrn(String,3,0);
   Sub3 = substrn(String,7,5);
   Sub4 = substrn(String,0,2);
   file print;
   put "Original string ="       @25 String   /
       "Substrn(string,2,5)  ="  @25 Result   /
       "Substrn(string,-1,4) ="  @25 Sub1     /
       "Substrn(string,3,0)  ="  @25 Sub2     /
       "Substrn(string,7,5)  ="  @25 Sub3     /
       "Substrn(string,0,2)  ="  @25 Sub4;
run;
```

Explanation

In data set HOAGIE (sub-strings, get it?) the storage lengths of the variables SUB1–SUB4 are all equal to the length of STRING (which is 10). Since a LENGTH statement was used to define the length of RESULT, it has a length of 5.

Examine the following results and the brief explanation that follows the results.

```
Demonstrating the SUBSTRN Function
Original string =        abcdefghij
Substrn(string,2,5) =    bcdef
Substrn(string,-1,4) =   ab
Substrn(string,3,0) =
Substrn(string,7,5) =    ghij
Substrn(string,0,2) =    a
```

The first use of the SUBSTRN function (RESULT) produces the same result as the SUBSTR function. The resulting substring starts at the second position in STRING (B) and has a length of 5. All the remaining SUBSTRN functions would have resulted in an error if the SUBSTR function had been used instead.

The starting position of –1 and a length of 4 results in the characters "AB". To figure this out, realize that you start counting from –1 (–1, 0, 1, 2) and the result is the first two characters in STRING.

When the LENGTH is 0, the result is a string of length 0.

When you start at position 7 and have a length of 5, you go past the end of the string, so that the result is truncated and the result is "GHIJ".

Finally, when the starting position is 0 and the length is 2, you get the first character in STRING, "A".

Function: **CHAR**

Purpose: To extract a single character from a string. What makes this function unique is that the default length of the result is 1 byte. Except for this feature, this function is similar to the SUBSTR (or SUBSTRN) function with the value of the third argument (the length) set to 1.

Syntax: CHAR(*character-value, position*)

character-value is any SAS character value.

position is the position of the character to be returned. If this value is 0 or negative or greater than the length of *character-value*, CHAR returns a missing value (with a length of 1). If position has a missing value, then CHAR returns a string with a length of 0.

Examples

For these examples, STRING = "ABC123".

Function	Returns
CHAR(STRING,2)	"B"
CHAR(STRING,7)	" "
CHAR(STRING,5)	"2"

Program 1.26: Demonstrating the CHAR function

```
***Primary function: CHAR;
***Other functions: LENGTHC;
data char;
   String = "ABC123";
   First = char(String,1);
   Length = lengthc(First);
   Second = char(String,2);
   Beyond = char(String,9);
   L_Beyond = lengthc(Beyond);
run;
title"Demonstrating the CHAR function";
proc print data=char noobs;
run;
```

Explanation

The main advantage of the CHAR function is that it saves you the trouble of using a
LENGTH statement to define the length of the result. Here is a listing of data set CHAR.
Notice that the storage length of results is 1.

```
Demonstrating the CHAR function

String    First    Length    Second    Beyond    L_Beyond

ABC123    A          1        B                      1
```

Function: FIRST

Purpose: To extract the first character from a string. This function is similar to the
CHAR function where the position argument is set to 1. Like the CHAR
function, the default length for the resulting value, if it has not been
previously defined, is equal to 1.

Syntax: `FIRST(character-value)`

character-value is any SAS character value.

If *character-value* is a missing value, the FIRST function returns a single blank.

Examples

For these examples, `STRING = "ABC123"`.

Function	Returns
FIRST(STRING)	"A"
FIRST("XYZ")	"X"
FIRST(" ")	" "

If you substitute FIRST(STRING) for CHAR(STRING,1) in the preceding program, you will obtain the same output.

Here is another example:

Program 1.27: Demonstrating the FIRST function

```
***Primary function: FIRST;
data names;
   input (First Middle Last)(: $10.);
   Initials = first(First) || first(Middle) || first(Last);
datalines;
Brian Page Watson
Sarah Ellen Washington
Nelson W. Edwards
;
title "Listing of data set NAMES";
proc print data=names noobs;
run;
```

Explanation

Although the SUBSTR or CHAR functions could have been used in this example, the FIRST function makes the solution quite elegant. Notice that a LENGTH statement is not needed here: the FIRST function returns a string of length 1, by default. Also, since the concatenation operator is used instead of one of the CAT functions, the length of the variable

INITIALS is 3, the sum of the lengths of each of the three arguments. (Remember that the CAT functions return a string of length 200 if the length of the resulting variable has not been previously set). Also note that in this example, everyone has a first name, middle name (or initial) and last name.

Here is the listing:

```
Listing of data set NAMES

First      Middle    Last          Initials

Brian      Page      Watson        BPW
Sarah      Ellen     Washington    SEW
Nelson     W.        Edwards       NWE
```

Functions That Join Two or More Strings Together

There are three CALL routines and five functions that concatenate character strings. Although you can use the concatenation operator (||) in combination with the STRIP, TRIM, or LEFT functions, these routines and functions make it much easier to put strings together and, if you want, to place one or more separator characters between the strings.

Besides stripping blanks and adding separators, there are two other advantages of using the "CAT" functions: one, you can supply a list of variables in the form BaseN-BaseM (allowing you to concatenate large numbers of values), and two, you can use **numeric** values as arguments to these functions (without any character-to-numeric conversion messages in the log).

The three CALL routines are discussed first, followed by the five concatenation functions.

CALL Routines

These three CALL routines concatenate two or more strings. Note that there are five concatenation functions as well (CAT, CATS, CATT, CATX, and CATQ). The differences among these routines involve the handling of leading and/or trailing blanks as well as spacing between the concatenated strings. The traditional concatenation operator (||) is still useful, but it sometimes takes extra work to strip leading and trailing blanks (LEFT and TRIM functions, or the new STRIP function) before performing the concatenation operation. These CALL routines are a convenience, and you will probably want to use them in place of the older form of concatenation. There are corresponding concatenation functions described in the next section.

One advantage of using the CALL routines over their corresponding functions is improved performance. For example, CALL CATS(R, X, Y, Z) is faster than R = CATS(R, X, Y, Z).

We will describe all three CALL routines and follow with one program demonstrating all three.

Function: CALL CATS

Purpose: To concatenate two or more strings, removing both leading and trailing blanks before the concatenation takes place. To help you remember that this CALL routine is the one that strips the leading and trailing blanks before concatenation, think of the S at the end of CATS as "strip blanks."

Note: To call our three cats, I usually just whistle loudly.

Syntax: `CALL CATS(result, string-1 <,string-n>)`

`result` is the concatenated string. It can be a new variable or, if it is an existing variable, the other strings will be added to it. **Be sure that the length of `result` is long enough to hold the concatenated results.** If not, the resulting string will be truncated, and you will see an error message in the log.

`string-1` and `string-n` are the character strings to be concatenated. Leading and trailing blanks will be stripped prior to the concatenation.

Example

For these examples,
```
A = "Bilbo" (no blanks)
B = "   Frodo" (leading blanks)
C = "Hobbit   " (trailing blanks)
D = "   Gandalf   " (leading and trailing blanks)
```

Note: `result` needs to be defined as a character variable (perhaps with a LENGTH statement) before the CALL routine.

Function	Returns
CALL CATS(RESULT, A, B)	"BilboFrodo"
CALL CATS(RESULT, B, C, D)	"FrodoHobbitGandalf"
CALL CATS(RESULT, "Hello", D)	"HelloGandalf"

Function: CALL CATT

Purpose: To concatenate two or more strings, removing only trailing blanks before the concatenation takes place. To help you remember this, think of the T at the end of CATT as "trailing blanks" or "trim blanks."

Syntax: **CALL CATT(*result*, *string-1* <,*string-n*>)**

result is the concatenated string. It can be a new variable or, if it is an existing variable, the other strings will be added to it. **Be sure that the length of *result* is long enough to hold the concatenated results.** If not, the program will terminate and you will see an error message in the log.

string-1 and *string-n* are the character strings to be concatenated. Trailing blanks will be stripped prior to the concatenation.

Example
For these examples,
A = "Bilbo" (no blanks)
B = "   Frodo" (leading blanks)
C = "Hobbit   " (trailing blanks)
D = "   Gandalf   " (leading and trailing blanks)

Note: *result* needs to be defined as a character variable (perhaps with a LENGTH statement) before the CALL routine.

Function	Returns
CALL CATT(RESULT, A, B)	"Bilbo Frodo"
CALL CATT(RESULT, B, C, D)	" FrodoHobbit Gandalf"
CALL CATT(RESULT, "Hello", D)	"Hello Gandalf"

Function: CALL CATX

Purpose: To concatenate two or more strings, removing both leading and trailing blanks before the concatenation takes place, and place a single space, or one or more characters of your choice, between each of the strings. To help you remember this, think of the X at the end of CATX as "add eXtra blank."

Syntax: CALL CATX(*separator, result, string-1 <,string-n>*)

separator is one or more characters, placed in single or double quotation marks, that you want to use to separate the strings.

result is the concatenated string. It can be a new variable or, if it is an existing variable, the other strings will be added to it. **Be sure that the length of *result* is long enough to hold the concatenated results.** If not, the program will terminate and you will see an error message in the log.

string-1 and *string-n* are the character strings to be concatenated. Leading and trailing blanks will be stripped prior to the concatenation and a single blank will be placed between the strings.

Example

For these examples,
A = "Bilbo" (no blanks)
B = "   Frodo" (leading blanks)
C = "Hobbit   " (trailing blanks)
D = "   Gandalf   " (leading and trailing blanks)

Note: *result* needs to be defined as a character variable (perhaps with a LENGTH statement) before the CALL routine.

Function	Returns
CALL CATX(" ", RESULT, A, B)	"Bilbo Frodo"
CALL CATX(",", RESULT, B, C, D)	"Frodo,Hobbit,Gandalf"
CALL CATX(":", RESULT, "Hello", D)	"Hello:Gandalf"
CALL CATX(", ", RESULT, "Hello", D)	"Hello, Gandalf"
CALL CATX("***", RESULT, A, B)	"Bilbo***Frodo"

Program 1.28: Demonstrating the three concatenation CALL routines

```
***Primary functions: CALL CATS, CALL CATT, CALL CATX;

data call_cat;
    String1 = "ABC";         * no spaces;
    String2 = "DEF   ";      * three trailing spaces;
    String3 = "   GHI";      * three leading spaces;
    String4 = "   JKL   ";   * three leading and trailing spaces;
    length Result1 - Result4 $ 20;
    call cats(Result1,String2,String4);
    call catt(Result2,String2,String1);
    call catx(" ",Result3,String1,String3);
    call catx(",",Result4,String3,String4);
run;
title "Listing of Data Set CALL_CAT";
proc print data=call_cat noobs;
run;
```

Explanation

The three concatenation CALL routines each perform concatenation operations. The CATS CALL routine strips leading and trailing blanks; the CATT CALL routine removes trailing blanks before performing the concatenation; the CATX CALL routine is similar to the CATS CALL routine except that it inserts a separator character (specified as the first argument) between each of the concatenated strings.

According to SAS documentation, the CALL routines are more efficient to use than the concatenation operator combined with the TRIM and LEFT functions. For example:

```
RESULT5 = TRIM(STRING2) || " " || TRIM(LEFT(STRING3));
```

```
RESULT5 would be: "DEF GHI"
```

Here is a listing of data set CALL_CAT:

```
Listing of Data Set CALL_CAT

String1  String2  String3  String4  Result1  Result2  Result3  Result4

 ABC      DEF       GHI      JKL     DEFJKL   DEFABC   ABC GHI  GHI,JKL
```

The "CAT" Functions (CAT, CATS, CATT, CATX, and CATQ)

These five concatenation functions are very similar to the concatenation CALL routines described previously. However, since they are functions and not CALL routines, you need to name the new character variable to be created on the left-hand side of the equal sign and the function, along with its arguments, on the right-hand side of the equal sign. As with the concatenation CALL routines, we will describe the five functions together and then use a single program to demonstrate them.

One very useful feature of all of these functions is that the items to be concatenated may be character or numeric. And, if you include numeric items, they are treated as if they were actually character values and no conversion messages are printed to the log.

Function: CAT

Purpose: To concatenate (join) two or more character strings or numeric values, leaving leading and/or trailing blanks unchanged. This function accomplishes the same task as the concatenation operator (||). However, it is important to realize that the CAT function and the concatenation operator differ in the default length of the result: the default length of the result when you use the operator is the sum of the lengths of the strings being concatenated; the default length of the result when you use the CAT function is 200. This is keeping with the default length for many of the SAS®9 functions.

Syntax: CAT(*item-1, item-2 <,item-n>*)

item-1, item-2 <,item-n> are the character strings or numeric values to be concatenated. These arguments can also be written as CAT(OF C1-C5) where C1-C5 represents a variable list.

Note: It is **very important to set the length of the resulting character string**, using a LENGTH statement or other method, before using any of the concatenation functions. Otherwise, the length of the resulting string will default to 200. When you use the concatenation operator, the length of the resulting string is the sum of the lengths of the values being concatenated.

Example

For these examples,

A = "Bilbo" (no blanks)
B = "   Frodo" (leading blanks)
C = "Hobbit   " (trailing blanks)
D = "   Gandalf   " (leading and trailing blanks)
C1-C5 are five character variables, with the values of 'A', 'B', 'C', 'D', and 'E'
respectively.

Function	Returns
CAT(A, B)	"Bilbo Frodo"
CAT(B, C, D)	" FrodoHobbit Gandalf "
CAT("Hello", D)	"Hello Gandalf "
CAT(OF C1-C5)	"ABCDE"
CAT(830,634,7229)	"8306347229" (no numeric-to-character conversion messages in the SAS log)

Function: CATS

Purpose: To concatenate (join) two or more character strings or numeric values, stripping both leading and trailing blanks on character values.

Syntax: CATS(*item-1, item-2 <,item-n>*)

item-1, item-2, and *item-n* are the character strings or numeric values to be concatenated. These arguments can also be written as CATS(OF C1-C5) where C1-C5 represents a variable list.

Note: It is **very important to set the length of the resulting character string**, using a LENGTH statement or other method, before calling any of the concatenation functions. Otherwise, the length of the resulting string will default to 200.

Example

For these examples,
A = "Bilbo" (no blanks)
B = "   Frodo" (leading blanks)
C = "Hobbit   " (trailing blanks)
D = "   Gandalf   " (leading and trailing blanks)
C1-C5 are five character variables, with the values of 'A', 'B', 'C', 'D', and 'E' respectively.

Function	Returns
CATS(A, B)	"BilboFrodo"
CATS(B, C, D)	"FrodoHobbitGandalf"
CATS("Hello", D)	"HelloGandalf"
CATS(OF C1-C5)	"ABCDE"
CATS(830,634,7229)	"8306347229" (no numeric-to-character conversion messages in the SAS log)

Function: **CATT**

Purpose: To concatenate (join) two or more character strings or numeric values, stripping only trailing blanks on character values.

Syntax: **CATT(*item-1, item-2 <,item-n>*)**

item1, item-2, and *item-n* are the character strings or numeric values to be concatenated. These arguments can also be written as CATT(OF C1-C5) where C1-C5 represents a variable list.

Note: It is **very important to set the length of the resulting character string**, using a LENGTH statement or other method, before calling any of the concatenation functions. Otherwise, the length of the resulting string will default to 200.

Example

For these examples,
A = "Bilbo" (no blanks)
B = "   Frodo" (leading blanks)
C = "Hobbit   " (trailing blanks)
D = "   Gandalf   " (leading and trailing blanks)
C1-C5 are five character variables, with the values of 'A', 'B', 'C', 'D', and 'E' respectively.

Function	Returns
CATT(A, B)	"Bilbo Frodo"
CATT(B, C, D)	" FrodoHobbit Gandalf"
CATT("Hello", D)	"Hello Gandalf"
CATT(OF C1-C5)	"ABCDE"
CATT(830,634,7229)	"8306347229" (no numeric-to-character conversion messages in the SAS log)

Function: **CATX**

Purpose: To concatenate (join) two or more character strings or numeric values, stripping both leading and trailing blanks on character values and inserting one or more separator characters between the strings.

Syntax: **CATX(*separator, item-1, item-2 <,item-n>*)**

separator is one or more characters, placed in single or double quotation marks, to be used as separators between the concatenated strings.

item-1, *item-2*, and *item-n* are the character strings or numeric values to be concatenated. These arguments can also be written as CATX(" ",OF C1-C5), where C1-C5 represents a variable list.

Note: It is **very important to set the length of the resulting character string**, using a LENGTH statement or other method, before calling any of the concatenation functions. Otherwise, the length of the resulting string will default to 200.

Example

For these examples,
A = "Bilbo" (no blanks)
B = "   Frodo" (leading blanks)
C = "Hobbit   " (trailing blanks)
D = "   Gandalf   " (leading and trailing blanks)
C1-C5 are five character variables, with the values of 'A', 'B', 'C', 'D', and 'E'
respectively.

Function	Returns
CATX(" ", A, B)	"Bilbo Frodo"
CATX(":"B, C, D)	"Frodo:Hobbit:Gandalf"
CATX("***", "Hello", D)	"Hello***Gandalf"
CATX("," ,OF C1-C5)	"A,B,C,D,E"
CATX,"-",830,257,8362)	"830-634-7229" (no numeric-to-character conversion messages in the SAS log)

Function: **CATQ**

Purpose: This function is similar to the CATX function, since it joins strings or numeric values and allows you to choose a separator between each of the strings. If any of the strings to be concatenated contains the delimiter, single or double quotes are placed around that string. Unlike CATX, CATQ does not automatically strip leading and trailing blanks unless you use the s modifier. In addition, with the a modifier, you can place quotes around all of the concatenated strings. As with the other "CAT" functions, the arguments can be character or numeric.

Syntax: CATQ(*modifiers,<delimiter,>,item-1,item-2 <,
item-n>*)

modifiers are character constants, variables, or expressions that affect the operation of the function. There are a large number of available modifiers, but only the more popular ones are shown here. For a complete list of these modifiers, please see Product Documentation in the Knowledge Base, available at http://support.sas.com/documentation.

modifier(s) can be in uppercase or lowercase.

The following *modifiers*, placed in single or double quotation marks, can be used with CATQ:

a	place quotes around each of the strings to be concatenated.
1	use single quotes.
2	use double quotes.
d	treat the next argument as a delimiter.
s	strip leading and trailing quotes.
m	add delimiters after the first item, even if the item is a missing value.

delimiter is one or more characters, placed in single or double quotation marks, to be used as separators between the concatenated strings. If you specify any delimiters, you must specify the d modifier (delimiter).

item-1, item-2, item-n are the character strings or numeric values to be concatenated. These arguments can also be written as CATQ(" ",OF C1-C5), where C1-C5 represents a variable list.

Note: It is **very important to set the length of the resulting character string**, using a LENGTH statement or other method, before calling any of the concatenation functions. Otherwise, the length of the resulting string will default to 200.

Example

For these examples,
```
A = "Single"
B = "Blank Between"
C = "  Lead-and-trail  "
```

Function	Returns
CATQ(" ",A,B)	Single "Blank Between"
CATQ("a",A,B)	"Single" "Blank Between"
CATQ("d",":",A,B)	Single:Blank Between
CATQ(" ",12,34,56)	12 34 56
CATQ("a",12,34,56)	"12" "34" "56"
CATQ("1a",12,34,56)	'12' '34' '56'
CATQ("a",A,C)	"Single" " Lead-and-trail "
CATQ("as",A,C)	"Single" "Lead-and-trail"

Program 1.29: Demonstrating the five concatenation functions

```
***Primary functions: CAT, CATS, CATT, CATX, CATQ;

title "Demonstrating the CAT functions";
data _null_;
   file print;
   String1 = "ABC";        * no spaces;
   String2 = "DEF   ";     * three trailing spaces;
   String3 = "   GHI";     * three leading spaces;
   String4 = "   JKL   ";  * three leading and trailing spaces;
   length Join1 - Join10 $ 20;
   Join1 = cat(String2,String3);
   Join2 = cats(String2,String4);
   join3 = cats(12,34,56);
   Join4 = catt(String2,String1);
   Join5 = catx(" ",String1,String3);
   Join6 = catx(",",String3,String4);
   Join7 = catx("-",908,782,6562);
   Join8 = catq(" ",String1,String3);
   Join9 = catq("a",String1,String3);
   Join10 = catq("as",String1,String3);
   S1 = ':' || String1 || ':';
   S2 = ':' || String2 || ':';
   S3 = ':' || String3 || ':';
   S4 = ':' || String4 || ':';
   put /"String 1 " S1/
   "String 2 " S2/
   "String 3 " S3/
   "String 4 " S4/
   Join1= /
   Join2= /
   Join3= /
   Join4= /
   Join5= /
   Join6= /
   Join7= /
   Join8= /
   Join9= /
   Join10= / ;
run;
```

Explanation

Notice that each of the STRING variables differs with respect to leading and trailing blanks.
The CAT function is identical to the || operator (with the exception of the default length of
the result). The CATS function removes both leading and trailing blanks and is equivalent to
TRIM(LEFT(STRING2)) || TRIM(LEFT(STRING4). The CATT function trims

only trailing blanks. The CATX functions remove leading and trailing blanks. The CATX function is just like the CATS function but adds one or more separator characters (specified as the first argument), between each of the strings to be joined. Notice that the CATQ function only strips leading trailing blanks if you use the s modifier and, if you want quotes around all of the concatenated strings, you need to include the a modifier.

The four STRING variables were concatenated with semicolons on both sides so that you can see the leading and/or trailing blanks in the listing. Inspection of the following listing will help make all this clear:

```
Demonstrating the CAT functions

String 1 :ABC:
String 2 :DEF   :
String 3 :   GHI:
String 4 :   JKL   :
Join1=DEF      GHI
Join2=DEFJKL
Join3=123456
Join4=DEFABC
Join5=ABC GHI
Join6=GHI,JKL
Join7=908-782-6562
Join8=ABC "   GHI"
Join9="ABC" "   GHI"
Join10="ABC" "GHI"
```

The "CAT" functions, in combination with functions such as FINDC or COUNTC (see a description of this function later in the chapter), provide for a very compact way to search for a particular value among several variables or to count the number of times a value occurs across several variables. For example, suppose you have a data set containing the variables Q1–Q10 where each of these variables is either a 'Y' or 'N'. If you wanted to know if any of the 10 values were equal to 'Y', you could use the following line:

```
AnyOne = findc(cat(of q1-q10),'y','i') gt 0;
```

If you wanted to count the number of Y's in these 10 variables, you could write:

```
Number = countc(cat(of q1-q10),'y','i');
```

Thanks to Mike Zdeb for this clever suggestion.

Functions That Remove Blanks from Strings

There are times when you want to remove blanks from the beginning or end of a character string. The two functions LEFT and RIGHT merely shift the characters to the beginning or the end of the string, respectively. The TRIM, TRIMN, and STRIP functions are useful when you want to concatenate strings (although the concatenation functions will do this for you).

LEFT and RIGHT

These two functions left- or right-align text. Remember that the length of a character variable will not change when you use these two functions. If there are leading blanks, the LEFT function will shift the first non-blank character to the first position and move the extra blanks to the end; if there are trailing blanks, the RIGHT function will shift the non-blank text to the right and move the extra blanks to the left.

Function: **LEFT**

Purpose: To left-align text values. A subtle but important point: LEFT doesn't remove the leading blanks; it moves them to the end of the string. Thus, it doesn't change the storage length of the variable, even when you assign the result of LEFT to a new variable. The LEFT function is particularly useful if values were read with the $CHAR informat, which preserves leading blanks. Note that the STRIP function removes both leading and trailing blanks from a string.

Syntax: `LEFT(character-value)`

   character-value is any SAS character value.

Example

In these examples, `STRING = "   ABC"`.

Function	Returns
`LEFT(STRING)`	`"ABC   "`
`LEFT("   123   ")`	`"123      "`

Program 1.30: Left-aligning text values from variables read with the $CHAR informat

```
***Primary function: LEFT;

data lead_on;
   input String $char15.;
   Left_string = left(String);
datalines;
ABC
   XYZ
  Ron Cody
;
title "Listing of Data Set LEAD_ON";
proc print data=lead_on noobs;
   format String Left_string $quote.;
run;
```

Explanation

If you want to work with character values, you will usually want to remove any leading blanks first. The $CHARw. informat differs from the $w. informat. $CHARw. maintains leading blanks; $w. left-aligns the text. Programs involving character variables sometimes fail to work properly because careful attention was not paid to either leading or trailing blanks. Of course, if you didn't want leading blanks and you were reading raw data, you would use the $15. informat instead of the $char15. informat—but we wanted to demonstrate how the LEFT function can left-justify character values.

Notice the use of the $QUOTE format in the PRINT procedure. This format adds double quotation marks around the character value. This is especially useful in debugging programs involving character variables, since it allows you to easily identify leading (but not trailing) blanks in a character value.

The listing of data set LEAD_ON follows. Notice that the original variable STRING contains leading blanks. The length of LEFT_STRING is also 15.

```
Listing of Data Set LEAD_ON


                Left_
String          string

"ABC"           "ABC"
"   XYZ"        "XYZ"
"  Ron Cody"    "Ron Cody"
```

Function: **RIGHT**

Purpose: To right-align a text string. Note that if the length of a character variable has previously been defined and it contains trailing blanks, the RIGHT function will move the characters to the end of the string and add the blanks to the beginning so that the final length of the variable remains the same.

Syntax: RIGHT(*character-value*)

character-value is any SAS character value.

Example

In these examples, STRING = "ABC   ".

Function	Returns
RIGHT(STRING)	" ABC"
RIGHT(" 123 ")	" 123"

Program 1.31: Right-aligning text values

```
***Primary function: RIGHT;

data right_on;
   input String $char10.;
   Right_string = right(String);
datalines;
   ABC
   123 456
Ron Cody
;
title "Listing of Data Set RIGHT_ON";
proc print data=right_on noobs;
   format String Right_string $quote.;
run;
```

Explanation

Data lines one and two both contain three leading blanks; lines one and three contain trailing blanks.

Notice the use of the $QUOTE format in the PRINT procedure. This format adds double quotation marks around the character value. This is especially useful in debugging programs involving character variables, since it allows you to easily identify leading (but not trailing) blanks in a character value.

Notice in the following listing that the values are right-aligned and that blanks are moved to the beginning of the string:

```
Listing of Data Set RIGHT_ON

String          Right_string

"    ABC"        "        ABC"
"    123 456"    "    123 456"
"Ron Cody"       "   Ron Cody"
```

TRIM, TRIMN, and STRIP

This group of functions trims trailing blanks (TRIM and TRIMN) and both leading and trailing blanks (STRIP).

The two functions TRIM and TRIMN are similar: they both remove trailing blanks from a string. The functions work identically except when the argument contains only blanks. In that case, TRIM returns a single blank (length of 1) and TRIMN returns a null string with a length of 0. The STRIP function removes both leading and trailing blanks.

Function: TRIM

Purpose: To remove trailing blanks from a character value. This is especially useful when you want to concatenate several strings together and each string may contain trailing blanks.

Syntax: TRIM(*character-value*)

character-value is any SAS character value.

Important note: The length of the variable returned by the TRIM function will be the same length as the argument, unless the length of this variable has been previously defined. If the result of the TRIM function is assigned to a variable with a length longer than the trimmed argument, the resulting variable will be padded with blanks.

Examples

For these examples, STRING1 = "ABC   " and STRING2 = "   XYZ".

Function	Returns
TRIM(STRING1)	"ABC"
TRIM(STRING2)	" XYZ"
TRIM("A B C ")	"A B C"
TRIM("A ") || TRIM("B ")	"AB"
TRIM(" ")	" " (length = 1)

Program 1.32: Creating a program to concatenate first, middle, and last names into a single variable

```
***Primary function: TRIM;

data put_together;
   length Name $ 45;
   informat Name1-Name3 $15.;
   infile datalines missover;
   input Name1 Name2 Name3;
   Name = trim(Name1) || ' ' || trim(Name2) || ' ' || Name3;
   Without = Name1 || Name2 || Name3;
   keep Name Without;
datalines;
Ronald Cody
Julia     Child
Henry     Ford
Lee Harvey Oswald
;
title "Listing Of Data Set PUT_TOGETHER";
proc print data=put_together noobs;
run;
```

Explanation

Note that this program would be much simpler using the concatenation functions or CALL routines. However, this method was used to demonstrate the TRIM function.

This program reads in three names, each up to 15 characters in length. Note the use of the INFILE option MISSOVER. This options sets the value of NAME3 to missing when there are only two names.

To put the names together, you use the concatenate operator (||). The TRIM function is used to trim trailing blanks from each of the words (which are all 15 bytes in length), before putting them together. Without the TRIM function, there are extra spaces between each of the names (see the variable WITHOUT). The following listing demonstrates that the program works as desired.

```
Listing Of Data Set PUT_TOGETHER

Name                Without

Ronald Cody         Ronald          Cody
Julia Child         Julia           Child
Henry Ford          Henry           Ford
Lee Harvey Oswald   Lee             Harvey          Oswald
```

Function: **TRIMN**

Purpose: To remove trailing blanks from a character value. This is especially useful when you want to concatenate several strings together and each string may contain trailing blanks. The difference between TRIM and TRIMN is that the TRIM function returns a single blank for a blank string, while TRIMN returns a null string (zero blanks).

Syntax: TRIMN(*character-value*)

character-value is any SAS character value.

Important note: The length of the variable returned by the TRIMN function will be the same length as the argument, unless the length of this variable has been previously defined. If the result of the TRIMN function is assigned to a variable with a length longer than the trimmed argument, the resulting variable will be padded with blanks.

Examples

For these examples, STRING1 = "ABC   " and STRING2 = "   XYZ".

Function	Returns
TRIMN(STRING1)	"ABC"
TRIMN(STRING2)	" XYZ"
TRIMN("A B C ")	"A B C"
TRIMN("A ") || TRIM("B ")	"AB"
TRIMN(" ")	"" (length = 0)

Program 1.33: Demonstrating the difference between the TRIM and TRIMN functions

```
***Primary functions: TRIM, TRIMN, and LENGTHC
***Other function: COMPRESS;

data all_the_trimmings;
   A = "AAA";
   B = "BBB";
   Length_ab = lengthc(A || B);
   Length_ab_trim = lengthc(trim(A) || trim(A));
   Length_ab_trimn = lengthc(trimn(A) || trimn(B));
   Length_null = lengthc(compress(A,"A") || compress(B, "B"));
   Length_null_trim = lengthc(trim(compress(A,"A")) ||
                     trim(compress(B,"B")));
   Length_null_trimn = lengthc(trimn(compress(A,"A")) ||
                     trimn(compress(B,"B")));
   put A= B= /
      Length_ab= Length_ab_trim= Length_ab_trimn= /
      Length_null= Length_null_trim= Length_null_trimn=;
run;
```

Explanation

First, remember that the LENGTHC function returns the length of its argument, including trailing blanks. As the following listing from the SAS log shows, the two functions TRIM and TRIMN yield identical results when there are no null strings involved. When you compress an "A" from the variable A, or "B" from variable B, the result is null. Notice that when you trim these compressed values and concatenate the results, the length is 2 (1 + 1); when you use the TRIMN function, the length is 0. Here are the lines written to the SAS log:

```
A=AAA B=BBB
Length_ab=6 Length_ab_trim=6 Length_ab_trimn=6
Length_null=0 Length_null_trim=2 Length_null_trimn=0
```

Function: **STRIP**

Purpose: To strip leading and trailing blanks from character variables or strings. STRIP(CHAR) is equivalent to TRIMN(LEFT(CHAR)), but more convenient.

Syntax: STRIP(*character-value*)

character-value is any SAS character value.

If the STRIP function is used to create a new variable, the length of that new variable will be equal to the length of the argument of the STRIP function. If leading or trailing blanks were trimmed, trailing blanks will be added to the result to pad out the length as necessary. The STRIP function is useful when using the concatenation operator. However, note that there are several new concatenation functions and CALL routines that also perform trimming before concatenation.

Examples

For these examples, STRING = "   abc   ".

Function	Returns
STRIP(STRING)	"abc" (if result was previously assigned a length of three, otherwise trailing blanks would be added)
STRIP(" LEADING AND TRAILING ")	"LEADING AND TRAILING"

Program 1.34: Using the STRIP function to strip both leading and trailing blanks from a string

```
***Primary function: STRIP;

data _null_;
   One = "   ONE   "; ***note: three leading and trailing blanks;
   Two = "   TWO   "; ***note: three leading and trailing blanks;
   Cat_no_strip = ":" || One || "-" || Two || ":";
   Cat_strip    = ":" || strip(One) || "-" || strip(Two) || ":";
   put One= Two= / Cat_no_strip= / Cat_strip=;
run;
```

Explanation

Without the STRIP function, the leading and trailing blanks are maintained in the concatenated string. The STRIP function, as advertised, removed the leading and trailing blanks. The following lines were written to the SAS log:

```
One=ONE Two=TWO
Cat_no_strip=:   ONE    -    TWO    :
Cat_strip=:ONE-TWO:
```

Functions That Compare Strings (Exact and "Fuzzy" Comparisons)

Functions in this section enable you to compare strings that are exactly alike (similar except for case) or close (not exact matches). Programmers find this latter group of functions useful in matching names that may be spelled differently in separate files.

Function: **COMPARE**

Purpose: To compare two character strings. When used with one or more modifiers, this function can ignore case, remove leading blanks, truncate the longer string to the length of the shorter string, and strip quotation marks from SAS n-literals. While all of these tasks can be accomplished with a variety of SAS functions, use of the COMPARE function can simplify character comparisons.

Syntax: COMPARE(*string-1, string-2 <,'modifiers'>*)

string-1 is any SAS character value.

string-2 is any SAS character value.

modifiers are one or more modifiers, placed in single or double quotation marks as follows:

        i or I   ignore case.

        l or L   remove leading blanks.

n or N   remove quotation marks from any argument that is an n-literal and ignore case.

An n-literal is a string in quotation marks, followed by an 'n', useful for non-valid SAS names.

: (colon)   truncate the longer string to the length of the shorter string. Note that the default is to pad the shorter string with blanks before a comparison; this is similar to the =: comparison operator.

Note that the order of the modifiers is important. See the following examples.

The function returns a 0 when the two strings match (after any modifiers are applied). If the two strings differ, a non-zero result is returned. The returned value is negative if *string-1* comes before *string-2* in a sort sequence, positive otherwise. The magnitude of this value is the position of the first difference in the two strings.

Examples

For these examples, string1 = "AbC", string2 = "   ABC", string3 = "   'ABC'n", string4 = "ABCXYZ".

Function	Returns
COMPARE(string1,string4)	2 ("B" comes before "b")
COMPARE(string4,string1)	-2
COMPARE(string1,string2,'i')	1
COMPARE(string1,string4,':I')	0
COMPARE(string1,string3,'nl')	4
COMPARE(string1,string3,'ln')	1

Program 1.35: Comparing two strings using the COMPARE function

```
***Primary function: COMPARE
***Other function: UPCASE;

data compare;
   input @1 String1 $char3.
         @5 String2 $char10.;
   if upcase(String1) = upcase(String2) then Equal = 'Yes';
   else Equal = 'No';
   if upcase(String1) =: upcase(String2) then Colon = 'Yes';
   else Colon = 'No';
   Compare = compare(String1,String2);
   Compare_il = compare(String1,String2,'il');
   Compare_il_colon = compare(String1,String2,'il:');
datalines;
Abc    ABC
abc ABCDEFGH
123 311
;
title "Listing of Data Set COMPARE";
proc print data=compare noobs;
run;
```

Explanation

The first two variables, EQUAL and COLON, use the UPCASE function to convert all the characters to uppercase before the comparison is made. The colon modifier following the equal sign (the variable COLON) is an instruction to truncate the longer variable to the length of the shorter variable before a comparison is made. Be careful here. If the variable STRING1 had been read with a $CHAR10. informat, the =: comparison operator would not have produced a match. The length truncation is done on the storage length of the variable, which may include trailing blanks.

The three COMPARE functions demonstrate the coding efficiency of using this function with its many modifiers. Use of this function without modifiers gives you no advantage over a simple test of equality using an equal sign. Using the il and colon modifiers enables you to compare the two strings, ignoring case, removing leading blanks, and truncating the two strings to a length of 3 (the length of STRING1). Note the value of COMPARE_IL_COLON in the third observation is –1, since "1" comes before "3" in the ASCII collating sequence. Here is the output from PROC PRINT:

```
Listing of Data Set COMPARE


                                              Compare_    Compare_
String1    String2    Equal    Colon   Compare    il      il_colon

   Abc        ABC       No       No        1        0          0
   abc     ABCDEFGH     No       Yes       1       -4          0
   123        311       No       No       -1       -1         -1
```

CALL COMPCOST, COMPGED, and COMPLEV

The two functions COMPGED and COMPLEV are both used to determine the similarity between two strings. The COMPCOST CALL routine enables you to customize the scoring system when you are using the COMPGED function.

COMPGED computes a quantity called **generalized edit distance**, which is useful in matching names that are not spelled exactly the same. The larger the value, the more dissimilar the two strings. COMPLEV performs a similar function but uses a method called the **Levenshtein edit distance**. It is more efficient than the generalized edit distance, but may not be as useful in name matching. See the SPEDIS function for a discussion of fuzzy merging and for detailed programming examples.

Function: CALL COMPCOST

Purpose: To determine the similarity between two strings, using a method called the generalized edit distance. The cost is computed based on the difference between the two strings. For example, replacing one letter with another is assigned one cost and reversing two letters can be assigned another cost. Since there is a default cost associated with every operation used by COMPGED, you can use that function without using COMPCOST at all. You need to call this function only once in a DATA step. Since this is a very advanced and complicated routine, only a few examples of its use will be explained. For a complete list of operations and costs, see Product Documentation in the Knowledge Base, available at http://support.sas.com/documentation.

Syntax: CALL COMPCOST('*operation-1*', *cost-1* <,'*operation-2*', *cost-2* ...>)

operation is a keyword, placed in quotation marks. A few keywords are listed here for explanation purposes, but see the product documentation for a complete list of operations:

Partial list of operations:
DELETE=
REPLACE=
SWAP=
TRUNCATE=

cost is a value associated with the operation. Valid values for cost range from –32,767 to +32,767.

Note: The wording of arguments in this book might differ from the wording of arguments in the product documentation.

Examples

```
CALL COMPCOST('REPLACE=', 100, 'SWAP=', 200);
CALL COMPCOST('SWAP=', 150);
```

Note: Operation can be uppercase or lowercase.

To see how CALL COMPCOST and COMPGED are used together, see Program 1.38.

Function: COMPGED

Purpose: To compute the similarity between two strings, using a method called the generalized edit distance. Since this function has similar uses to the SPEDIS function, please refer to the discussion and examples of the SPEDIS function to understand how you might use the COMPGED function.

This function can be used in conjunction with CALL COMPCOST if you want to alter the default costs for each type of spelling error.

Syntax: COMPGED(*string-1, string-2 <,maxcost>*
<,*'modifiers'*>)

string-1 is any SAS character value.

string-2 is any SAS character value.

maxcost, if specified, is the maximum cost that will be returned by the
COMPLEV function. If the cost computation results in a value larger than
maxcost, the value of *maxcost* will be returned.

The following *modifiers,* placed in single or double quotation marks,
can be used with COMPGED:

i or I	ignore case.
l or L	remove leading blanks.
n or N	remove quotation marks from any argument that is an n-literal and ignore case.
	An n-literal is a string in quotation marks, followed by an 'n', useful for non-valid SAS names.
: (colon)	truncate the longer string to the length of the shorter string.

Note: If multiple modifiers are used, the order of the modifiers is important.
They are applied in the same order as they appear.

Examples

String1	String2	Function	Returns
SAME	SAME	COMPGED(STRING1, STRING2)	0
case	CASE	COMPGED(STRING1, STRING2)	500
case	CASE	COMPGED(STRING1,STRING2,'I')	0
case	CASE	COMPGED(STRING1, STRING2, 999, 'I')	0
Ron	Run	COMPGED(STRING1, STRING2)	100

Program 1.36: Using the COMPGED function with a SAS n-literal

```
***Primary function: COMPGED;

options validvarname=any;

data n_literal;
    String1 = "'Invalid#'n";
    String2 = 'Invalid';
    Comp1 = compged(String1,String2);
    Comp2 = compged(String1,String2,'n:');
run;

title "Listing of Data Set N_LITERAL";
proc print data=n_literal noobs;
run;
```

Explanation

This program demonstrates the use of the COMPGED function with a SAS n-literal. Starting with SAS 7, SAS variable names could contain characters not normally allowed in SAS names. The system option VALIDVARNAME is set to ANY, and the name is placed in quotation marks, followed by the letter N. Using the n modifier (which strips quotation marks and the 'n' from the string) and the colon modifier (which truncates the longer string to the length of the shorter string) results in a value of 0 for the variable COMP2. See the following listing:

```
Listing of Data Set N_LITERAL

  String1      String2   Comp1   Comp2

'Invalid#'n    Invalid    310       0
```

Function: COMPLEV

Purpose: To compute the similarity between two strings, using a method called the **Levenshtein edit distance**. It is similar to the COMPGED function except that it uses less computer resources but may not do as good a job of matching misspelled names.

Syntax: COMPLEV(*string-1, string-2 <,maxcost>* *<,'modifiers'>*)

string-1 is any SAS character value.

string-2 is any SAS character value.

maxcost, if specified, is the maximum cost that will be returned by the COMPGED function. If the cost computation results in a value larger than *maxcost*, the value of *maxcost* will be returned.

modifiers (placed in single or double quotation marks):

i or I	ignore case.
l or L	remove leading blanks.
n or N	remove quotation marks from any argument that is an n-literal and ignore case.
	An n-literal is a string in quotation marks, followed by an 'n', that is useful for non-valid SAS names.
: (colon)	truncate the longer string to the length of the shorter string.

Note: If multiple modifiers are used, the order of the modifiers is important. They are applied in the same order as they appear.

Examples

String1	String2	Function	Returns
SAME	SAME	COMPLEV(STRING1, STRING2)	0
case	CASE	COMPLEV(STRING1, STRING2)	4
case	CASE	COMPLEV(STRING1,STRING2,'I')	0
case	CASE	COMPLEV(STRING1, STRING2, 999, 'I')	0
Ron	Run	COMPLEV(STRING1, STRING2)	1

Program 1.37: Demonstration of the generalized edit distance (COMPGED) and Levenshtein edit distance (COMPLEV) functions

```
***Primary functions: COMPGED and COMPLEV;

title "Demonstrating COMPGED and COMPLEV functions";
data _null_;
   input @1  String1 $char10.
         @11 String2 $char10.;
   file print;
   put "Function COMPGED";
   Distance = compged(String1, String2);
   Ignore_case = compged(String1, String2, 'i');
   Lead_blanks = compged(String1, String2, 'l');
   Case_trunc = compged(String1, String2, ':i');
   Max = compged(String1, String2, 250);
   put String1= String2= /
       Distance= Ignore_case= Lead_blanks= Case_trunc= Max= /;

   put "Function COMPLEV";
   Distance = complev(String1, String2);
   Ignore_case = complev(String1, String2, 'i');
   Lead_blanks = complev(String1, String2, 'l');
   Case_trunc = complev(String1, String2, ':i');
   Max = complev(String1, String2, 3);
   put String1= String2= /
       Distance= Ignore_case= Lead_blanks= Case_trunc= Max= /;
datalines;
SAME      SAME
cAsE      case
Longer    Long
abcdef    xyz
   lead   lead
;
```

Explanation

In this program, all the default costs were used, so it was not necessary to call COMPCOST. Notice that the strings were read in with the $CHAR. informat so that leading blanks would be preserved. If you want to use modifiers, you must enter them in quotation marks, in the order you want the modifying operations to proceed. Careful inspection of the following SAS log demonstrates the COMPGED and COMPLEV functions and the effects of the modifiers and the *maxcost* parameter.

Output from Program 1.37

```
Demonstrating COMPGED and COMPLEV functions

Function COMPGED
String1=SAME String2=SAME
Distance=0 Ignore_case=0 Lead_blanks=0 Case_trunc=0 Max=0

Function COMPLEV
String1=SAME String2=SAME
Distance=0 Ignore_case=0 Lead_blanks=0 Case_trunc=0 Max=0

Function COMPGED
String1=cAsE String2=case
Distance=200 Ignore_case=0 Lead_blanks=200 Case_trunc=0 Max=200

Function COMPLEV
String1=cAsE String2=case
Distance=2 Ignore_case=0 Lead_blanks=2 Case_trunc=0 Max=2

Function COMPGED
String1=Longer String2=Long
Distance=100 Ignore_case=100 Lead_blanks=100 Case_trunc=100 Max=100

Function COMPLEV
String1=Longer String2=Long
Distance=2 Ignore_case=2 Lead_blanks=2 Case_trunc=2 Max=2

Function COMPGED
String1=abcdef String2=xyz
Distance=550 Ignore_case=550 Lead_blanks=550 Case_trunc=550 Max=250

Function COMPLEV
String1=abcdef String2=xyz
Distance=6 Ignore_case=6 Lead_blanks=6 Case_trunc=6 Max=3

Function COMPGED
String1=lead String2=lead
Distance=320 Ignore_case=320 Lead_blanks=0 Case_trunc=320 Max=250

Function COMPLEV
String1=lead String2=lead
Distance=3 Ignore_case=3 Lead_blanks=0 Case_trunc=3 Max=3
```

The following program demonstrates how to use CALL COMPCOST in combination with COMPGED and the resulting differences.

Program 1.38: Changing the effect of the call to COMPCOST on the result from COMPGED

```
***Primary functions: CALL COMPCOST and COMPGED;

title "Program without Call to COMPCOST";
data _null_;
   input @1  String1 $char10.
         @11 String2 $char10.;
   Distance = compged(String1, String2);
   put String1= String2= /
       Distance=;
datalines;
Ron        Run
ABC        AB
;

title "Program with Call to COMPCOST";
data _null_;
   input @1  String1 $char10.
         @11 String2 $char10.;
   if _n_ = 1 then call compcost('append=',33);
   Distance = compged(String1, String2);
   put String1= String2= /
       Distance=;
datalines;
Ron        Run
ABC        AB
;
```

Explanation

The first DATA _NULL_ program is a simple comparison of STRING1 to STRING2, using the COMPGED function. The second DATA _NULL_ program makes a call to COMPCOST (note the use of _N_ = 1) before the COMPGED function is used. In the following SAS logs, you can see that the distance in the second observation in the first program is 50, while in the second program it is 33. That is the result of overriding the default value of 50 points for an appending error and setting it equal to 33.

SAS Log from Program without CALL COMPCOST

```
String1=Ron String2=Run
Distance=100
String1=ABC String2=AB
Distance=50
```

SAS Log from Program with CALL COMPCOST

```
String1=Ron String2=Run
Distance=100
String1=ABC String2=AB
Distance=33
```

Function: SOUNDEX

The SOUNDEX function creates a phonetic equivalent of a text string to facilitate fuzzy matching. You can research the details of the SOUNDEX algorithm in the Product Documentation in the Knowledge Base, available at http://support.sas.com/documentation. Briefly, this algorithm makes all vowels equal, along with letters that sound the same (such as C and K). One feature of this algorithm is that the first letters in both words must be the same to obtain a phonetic match.

For those readers interested in the topic of fuzzy matching, there is an algorithm called NYSIIS, similar to the SOUNDEX algorithm, that maintains vowel position information and allows mismatches on the initial letter. A copy of a macro to implement the NYSIIS algorithm is available on the author pages for this book, located at http://support.sas.com/authors. (Select Ron Cody from the list of authors, find the Functions book, and then click the link Example Code and Data. The NYSSIS macro is included as one of the files you can download.)

In addition, see the SPEDIS, COMPGED, and COMPLEV functions, as well as the COMPCOST CALL routine for alternative matching algorithms.

Purpose: To create a phonetic equivalent of a text string. Often used to attempt to match names where there might be some minor spelling differences.

Syntax: SOUNDEX(*character-value*)

character-value is any SAS character value.

Program 1.39: Fuzzy matching on names using the SOUNDEX function

```
***Primary function: SOUNDEX
***Prepare data sets FILE_1 and FILE_2 to be used in the match;

data file_1;
   input @1  Name  $10.
         @11 X          1.;
   Sound_1 = soundex(Name);
datalines;
Friedman   4
Shields    1
MacArthur  7
ADAMS      9
Jose       5
Lundquist  9
;
data file_2;
   input @1  Name  $10.
         @11 Y          1.;
   Sound_2 = soundex(Name);
datalines;
Freedman   5
Freidman   9
Schields   2
McArthur   7
Adams      3
Jones      6
Londquest  9
;
***PROC SQL is used to create a Cartesian product, combinations of all
the names in one data set against all the names in the other.

   If you are not familiar with a Cartesian product, run the
   following PROC SQL code without the WHERE clause.  Note that
   the size of the resulting table is the number of rows in
   one table times the number of rows in the other table.  With
   "real" data sets the resulting table would most likely be very
   large.;

proc sql;
   create table possible as
   select file_1.name as Name1, X,
          file_2.name as Name2, Y
   from file_1 ,file_2
   where Sound_1 eq Sound_2;
quit;

title "Possible Matches between two files";
proc print data=possible noobs;
run;
```

Explanation

Each of the two SAS data sets (FILE_1 and FILE_2) contain names, data, and the SOUNDEX value of the name. One way to compare every name in one file against every name in another, is to create what is called a Cartesian product. Since there are six observations in FILE_1 and seven observations in FILE_2, the Cartesian product would contain 6 x 7 = 42 observations. To see exactly how this works, the following listing shows the Cartesian product of data sets FILE_1 and FILE_2 (without the subsetting WHERE clause).

```
Possible Matches between two files

  Name1      X    Name2      Y    Sound_1    Sound_2

Friedman     4    Freedman   5    F6355      F6355
Friedman     4    Freidman   9    F6355      F6355
Friedman     4    Schields   2    F6355      S432
Friedman     4    McArthur   7    F6355      M2636
Friedman     4    Adams      3    F6355      A352
Friedman     4    Jones      6    F6355      J52
Friedman     4    Londquest  9    F6355      L53223
Shields      1    Freedman   5    S432       F6355
Shields      1    Freidman   9    S432       F6355
Shields      1    Schields   2    S432       S432
Shields      1    McArthur   7    S432       M2636
Shields      1    Adams      3    S432       A352
Shields      1    Jones      6    S432       J52
Shields      1    Londquest  9    S432       L53223
MacArthur    7    Freedman   5    M2636      F6355
```

In practice, this data set is never created (since it would probably be too large). Instead, a subsetting WHERE clause restricts the resulting data set to observations where the SOUNDEX of the names are equal.

All that you need to do now is to select all the observations (rows in SQLese) where the SOUNDEX equivalents of the names are equal. Here is the result:

```
Possible Matches between two files

Name1       X    Name2      Y

Friedman    4    Freedman   5
Friedman    4    Freidman   9
Shields     1    Schields   2
MacArthur   7    McArthur   7
ADAMS       9    Adams      3
Lundquist   9    Londquest  9
```

Function: SPEDIS

The SPEDIS function computes a "spelling distance" between two words. If the two words are identical, the spelling distance is 0; for each type of spelling error, SPEDIS assigns penalty points. For example, if the first letters of the two words do not match, there is a relatively large penalty. If two letters are reversed (such as 'ie' instead of 'ei'), there is a smaller penalty. The final spelling distance is also based on the length of the words being matched. A wrong letter in a long word results in a smaller score than a wrong letter in a shorter word.

Take a look at two new functions (COMPGED and COMPLEV) that are similar to the SPEDIS function. The COMPGED function even has an associated CALL routine (COMPCOST) that enables you to adjust the costs for various classes of spelling errors.

There are some very interesting applications of the SPEDIS, COMPGED, and COMPLEV functions. One, described next, enables you to match names that are not spelled exactly the same. You may want to try this same program with each of the alternative functions to compare matching ability and computer efficiency.

Purpose:

To compute the spelling distance between two words. The more alike the two words are, the lower the score. A score of 0 indicates an exact match.

Syntax: SPEDIS(*word-1, word-2*)

word-1 is any SAS character value.
word-2 is any SAS character value.

The function returns the spelling distance between the two words. A zero indicates the words are identical. Higher scores indicate the words are more dissimilar. Note that SPEDIS is asymmetric; the order of the values being compared makes a difference. The value returned by the function is computed as a percentage of the length of the string to be compared. Thus, an error in two short words returns a larger value than the same error in two longer words.

A list of the operations and their cost is shown in the following table. This will give you an idea of the severity of different types of spelling errors. Swapping (interchanging) two letters has a smaller penalty than inserting a letter, and errors involving the first character in a string have a larger penalty than errors involving letters in the middle of the string.

Operation	Cost	Explanation
MATCH	0	no change
SINGLET	25	delete one of a double letter
DOUBLET	50	double a letter
SWAP	50	reverse the order of two consecutive letters
TRUNCATE	50	delete a letter from the end
APPEND	35	add a letter to the end
DELETE	50	delete a letter from the middle
INSERT	100	insert a letter in the middle
REPLACE	100	replace a letter in the middle
FIRSTDEL	100	delete the first letter
FIRSTINS	200	insert a letter at the beginning
FIRSTREP	200	replace the first letter

Examples

For these examples, WORD1="Steven", WORD2 = "Stephen", and WORD3 = "STEVEN".

Function	Returns
SPEDIS(WORD1,WORD2)	25
SPEDIS(WORD2,WORD1)	28
SPEDIS(WORD1,WORD3)	83
SPEDIS(WORD1,"Steven")	0

Program 1.40: Using the SPEDIS function to match Social Security numbers that are the same or differ by a small amount

```
***Primary function: SPEDIS;

data first;
    input ID_1 : $2. SS : $11.;
datalines;
1A 123-45-6789
2A 111-45-7654
3A 999-99-9999
4A 222-33-4567
;
data second;
    input ID_2 : $2. SS : $11.;
datalines;
1B 123-45-6789
2B 111-44-7654
3B 899-99-9999
4B 989-99-9999
5B 222-22-5467
;
%let Cutoff = 10;
title "Output from SQL when CUTOFF is set to &CUTOFF";
proc sql;
    select ID_1,
           First.SS as First_SS,
           ID_2,
           Second.SS as Second_SS
    from first, second
    where spedis(First.SS, Second.SS) le &Cutoff;
quit;
```

Explanation

In this example, you want to match Social Security numbers between two files, where the two numbers may be slightly different (one digit changed or two digits transposed for example). Even though Social Security numbers are made up of digits, you can treat the values just like any other alphabetic characters. The program is first run with the CUTOFF value set to 10, with the result shown next:

```
Output from SQL when CUTOFF is set to 10

ID_1  First_SS    ID_2  Second_SS

1A    123-45-6789 1B    123-45-6789
2A    111-45-7654 2B    111-44-7654
3A    999-99-9999 4B    989-99-9999
```

When you run the same program with the CUTOFF value set to 20, you obtain an additional match—numbers 3A and 3B. These two numbers differ in the first digit, which resulted in a larger penalty. Here is the listing with CUTOFF set to 20:

```
Output from SQL when CUTOFF is set to 20

ID_1  First_SS    ID_2  Second_SS

1A    123-45-6789 1B    123-45-6789
2A    111-45-7654 2B    111-44-7654
3A    999-99-9999 3B    899-99-9999
3A    999-99-9999 4B    989-99-9999
```

Program 1.41: Fuzzy matching on names using the spelling distance (SPEDIS) function

```
***Primary function: SPEDIS
***Other function: PROPCASE;

data file_1;
   input @1  Name  $10.
         @11 X        1.;
datalines;
Friedman  4
Shields   1
MacArthur 7
ADAMS     9
Jose      5
Lundquist 9
;
```

```
data file_2;
   input @1  Name  $10.
         @11 Y       1.;
datalines;
Freedman   5
Freidman   9
Schields   2
McArthur   7
Adams      3
Jones      6
Londquest 9
;
***PROC SQL is used to create a Cartesian product, combinations of all
the names in one data set against all the names in the other.;

proc sql;
   create table possible as
   select file_1.name as Name1, X,
          file_2.name as Name2, Y
   from file_1 ,file_2
   where 0 lt spedis(propcase(Name1),propcase(Name2)) le 15;
quit;

title "Possible Matches between two files";
proc print data=possible noobs;
run;
```

Explanation

This program is similar to the previous (SOUNDEX) example. Instead of creating a new variable in each data set that is the SOUNDEX value of the name, this program uses PROC SQL to create the Cartesian product (the combination of every observation in one data set with every observation in the second data set). The WHERE clause tests whether the spelling distance between the two names is less than or equal to 15. In addition, exact matches are removed by specifying that the spelling distance must be greater than zero. Also, the PROPCASE function is used in case there are case differences between the two names being matched. (Note: the COMPGED and COMPLEV functions allow a modifier to ignore case in the comparison.)

As you can see from the following listing, this method of fuzzy matching produces a different list of matches from the one produced using the SOUNDEX function.

```
Possible Matches between two files

  Name1      X      Name2      Y

Friedman     4    Freedman     5
Friedman     4    Freidman     9
Shields      1    Schields     2
MacArthur    7    McArthur     7
```

Functions That Divide Strings into "Words"

These extremely useful functions and CALL routines can divide a string into words. Words can be characters separated by blanks or other delimiters that you specify.

SCAN

The SCAN function is the premier function that can separate parts of a character value. The most popular use of this function, at least by this author, is to obtain first, middle, and last names when they are all stored in a single variable (don't you just hate that?). With the ability to define delimiters and with the addition of modifiers (SAS 9.2), this function is even more powerful.

The first edition of this book included a discussion of the SCANQ function. This function has been dropped from this edition since all the features of SCANQ can be replicated by using the new (SAS 9.2) modifiers added to the SCAN function (especially the q and s modifiers).

Function: SCAN

Purpose: Extracts a specified word from a character value, where **word** is defined as the characters separated by a set of specified delimiters. The length of the returned variable is 200, unless previously defined.

Syntax: SCAN(*char-value, n <,'delimiters'<,'modifiers'>>*)

char-value is any SAS character value.

n is the *n*th "word" in the string. If *n* is greater than the number of words, the SCAN function returns a missing value. If *n* is negative, the character value is scanned from right to left. A value of zero is invalid.

delimiters is an optional argument. If it is omitted, the default delimiters for ASCII environments (used on PCs and UNIX machines) are as follows:

blank . < (+ & ! $ *) ; ^ - / , % |

For EBCDIC environments (used mainly on main-frame computers), the default delimiters are as follows:

blank . < (+ | & ! $ *) ; ¬ - / , % | ¢

If you specify any delimiters, only those delimiters will be active. Delimiters before the first word have no effect. Two or more contiguous delimiters are treated as one.

modifiers (added in SAS 9.2) is an optional list of predefined characters that either add character classes (such as all digits) to the list of delimiters or reverse the way that delimiters are interpreted (see specifically, the k modifier). If you want to specify modifiers, you must also supply a list of delimiters. With the exception of the k and i modifiers, the character classes specified by the modifiers are interpreted as delimiters in addition to the ones you supplied in the list of delimiters.

You can also use the m modifier to change the way that successive delimiters and delimiters at the beginning or end of the string are interpreted. Without the m modifier, multiple successive delimiters are treated as a single delimiter, and delimiters at the beginning or end of the string are ignored. Using the m modifier treats successive delimiters as separating strings of zero length. For more details on the m and other modifiers, see Product Documentation in the Knowledge Base, available at http://support.sas.com/documentation.

Note that modifiers were added in SAS 9.2.

A partial list of modifiers follows:

a	include all uppercase and lowercase letters.
c	include control characters.
d	include all digits.
h	include horizontal tabs.
i	ignore case.
k	use characters NOT in the delimiter list as delimiters.
l	include lowercase characters.
m	treat multiple successive delimiters differently (see the preceding explanation).
p	include all punctuation.
s	include all space characters (blanks, tabs, vertical and horizontal tabs).
u	include uppercase characters.

Examples

For these examples,

```
STRING1 = "ABC DEF"
STRING2 = "ONE?TWO THREE+FOUR|FIVE"
STRING3 = "abc1def2ghi"
STRING4 = "123,,456"
```

This is an ASCII example.

Function	Returns
SCAN(STRING1,2)	"DEF"
SCAN(STRING1,-1)	"DEF"
SCAN(STRING1,3)	missing value
SCAN(STRING2,4)	"FIVE" (? is not a default delimiter)
SCAN(STRING2,2," ")	"THREE+FOUR|FIVE"
SCAN(STRING1,0)	an error in the SAS log
SCAN(STRING3,2," ","d")	"def"
SCAN(STRING3,2," ","ka")	"def"
SCAN(STRING4,2," ,")	"456"
SCAN(STRING4,2," ,","m")	missing value (2 commas treated as two delimiters)

Program 1.42: A novel use of the SCAN function to convert mixed numbers to decimal values

```
***Primary function: SCAN
***Other function: INPUT;

data prices;
   input @1 Stock $3.
         @5 Mixed $6.;
   Integer = scan(Mixed,1,'/ ');
   Numerator = scan(Mixed,2,'/ ');
   Denominator = scan(Mixed,3,'/ ');
   if missing(Numerator) then Value = input(Integer,8.);
   else Value = input(Integer,8.) +
                  (input(Numerator,8.) / input(Denominator,8.));
   keep Stock Mixed Value;
datalines;
ABC 14 3/8
XYZ 8
TWW 5 1/8
;
title "Listing of Data Set PRICES";
proc print data=prices noobs;
run;
```

Explanation

The SCAN function has many uses besides merely extracting selected words from text expressions. In this program, you want to convert numbers such as 23 5/8 into a decimal value (23.675). An elegant way to accomplish this is to use the SCAN function to separate the mixed number into three parts: the integer, the numerator of the fraction, and the denominator. Once this is done, all you need to do is to convert each piece to a numerical value (using the INPUT function) and add the integer portion to the fractional portion. If the number being processed does not have a fractional part, the SCAN function returns a missing value for the two variables NUMERATOR and DENOMINATOR. Here is the listing:

```
Listing of Data Set PRICES

Stock    Mixed      Value

 ABC     14 3/8     14.375
 XYZ     8           8.000
 TWW     5 1/8       5.125
```

Program 1.43: Program to read a tab-delimited file

```
***Primary function: SCAN;

data read_tabs;
   infile 'c:\books\functions\tab_file.txt' pad;
   input @1 String $30.;
   length First Middle Last $ 12;
   First = scan(String,1,' ','h');
   Middle = scan(String,2,' ','h');
   Last = scan(String,3,' ','h');
   drop String;
run;
title "Listing of Data Set READS_TABS";
proc print data=read_tabs noobs;
run;
```

Explanation

This program reads values separated by tab characters. Although you can use the INFILE option DLM='09'X (the ASCII hexadecimal value for a tab character, or '05'X for EBCDIC) to read this file, the SCAN function with the h modifier (horizontal tabs) provides an easy, alternate method. This method could be especially useful if you imported a file from another system, and individual character values contained tabs or other non-printing white space characters.

```
Listing of Data Set READS_TABS

First     Middle     Last

Ron       P.         Cody
Ralph     Waldo      Emerson
Alfred    E.         Newman
```

Program 1.44: Alphabetical listing by last name when the name field contains first name, possibly middle initial, and last name

```
***Primary function: SCAN;

data first_last;
   input @1  Name  $20.
         @21 Phone $13.;
   ***extract the last name from name;
   Last_name = scan(Name,-1,' '); /* scans from the right */
datalines;
Jeff W. Snoker       (908)782-4382
```

```
Raymond Albert      (732)235-4444
Steven J. Foster    (201)567-9876
Jose Romerez        (516)593-2377
;
title "Names and Phone Numbers in Alphabetical Order (by Last Name)";
proc report data=first_last nowd;
   columns Name Phone Last_name;
   define Last_name / order noprint width=20;
   define Name       / display 'Name' left width=20;
   define Phone      / display 'Phone Number' width=13 format=$13.;
run;
```

Explanation

It is easy to extract the last name by using a −1 as the second argument of the SCAN function. A negative value for this argument results in a scan from right to left. Here is the output from the REPORT procedure:

```
Names and Phone Numbers in Alphabetical Order (by Last Name)

Name                Phone Number
Raymond Albert      (732)235-4444
Steven J. Foster    (201)567-9876
Jose Romerez        (516)593-2377
Jeff W. Snoker      (908)782-4382
```

CALL SCAN

The SCAN CALL routine is similar to the SCAN function. The CALL routine returns a position and length of the *n*th word (to be used, perhaps, in a subsequent SUBSTR function) rather than the actual word itself.

Function: CALL SCAN

Purpose: To break up a string into words, where **words** are defined as the characters separated by a set of specified delimiters, and to return the starting position and the length of the *n*th word.

Starting with SAS 9.2, the CALL routine allows you to supply modifiers (the same ones used with the function).

Syntax: CALL SCAN(*char-value, n, position, length*
<,*'delim'*> <,*'modifiers'*>)

char-value is any SAS character value.

n is the *n*th "word" in the string. If *n* is greater than the number of words, the SCAN CALL routine returns a value of 0 for *position* and *length*. If *n* is negative, the scan proceeds from right to left.

position is the name of the numeric variable to which the starting position in the *char-value* of the *n*th word is returned.

length is the name of a numeric variable to which the length of the *n*th word is returned.

delim is an optional argument that represents a character or list of characters that are to be treated as delimiters. If it is omitted, the default delimiters for ASCII environments are as follows:

   blank . < (+ & ! $ *) ; ^ - / , % |

For EBCDIC environments, the default delimiters are as follows:

   blank . < (+ | & ! $ *) ; ¬ - / , % | ¢

If you specify any delimiters, only those delimiters will be active. Delimiters are slightly different in ASCII and EBCDIC systems.

modifiers is a character value that specifies one or more modifiers. The list of modifiers and their meaning is identical to the ones used with the SCAN function.

Some of the modifiers simply add to the delimiter list (such as all lowercase letters). Some of the modifiers change the way the function works. For example, the m modifier specifies that multiple delimiters plus delimiters at the beginning or end of the string are to be treated as individual delimiters.

Examples

For these examples, STRING1 = "ABC DEF GHI" and
STRING2 = "ONE?TWO THREE+FOUR|FIVE".

Function	Position	Length
CALL SCAN(STRING1,2,POSITION,LENGTH)	5	3
CALL SCAN(STRING1,-1,POSITION,LENGTH)	9	3
CALL SCAN(STRING1,3,POSITION,LENGTH)	0	0
CALL SCAN(STRING2,1,POSITION,LENGTH)	1	7
CALL SCAN(STRING2,4,POSITION,LENGTH)	20	4
CALL SCAN(STRING2,2,POSITION,LENGTH," ")	9	15
CALL SCAN(STRING1,0,POSITION,LENGTH)	missing	missing

Program 1.45: Demonstrating the SCAN CALL routine

```
***Primary function: CALL SCAN;

data words;
   input String $40.;
   Delim = 'Default';
   N = 2;
   call scan(String,N,Position,Length);
   output;
   N = -1;
   call scan(String,N,Position,Length);
   output;
   Delim = '#';
   N = 2;
   call scan(String,N,Position,Length,'#');
   output;
datalines;
ONE TWO THREE
One*#Two Three*Four
;
title "Listing of Data Set WORDS";
proc print data=words noobs;
run;
```

Explanation

The SCAN routine is called three times in this program, twice with default delimiters and once with the pound sign (#) as the delimiter. Notice that using a negative argument results in a scan from right to left. The output from this program is shown next:

```
Listing of Data Set WORDS

        String              Delim      N    Position    Length

ONE TWO THREE               Default    2       5          3
ONE TWO THREE               Default   -1       9          5
ONE TWO THREE               #          2       0          0
One*#Two Three*Four         Default    2       5          4
One*#Two Three*Four         Default   -1      16          4
One*#Two Three*Four         #          2       6         35
```

Program 1.46: Using CALL SCAN to count the words in a string

```
***Primary function: CALL SCAN;

data count;
   input String $40.;
   do i = 1 to 99 until (Length eq 0);
      call scan(String,i,Position,Length);
   end;
   Num_words = i-1;
   drop Position Length i;
datalines;
ONE TWO THREE
ONE TWO
ONE
;
title "Listing of Data Set COUNT";
proc print data=count noobs;
run;
```

Explanation

Note that the COUNTW function will accomplish this task directly—we are using CALL SCAN here as an illustration of how this CALL routine works.

When the value of the second argument in the CALL SCAN routine is greater than the number of words in a string, both the position and length values are set to 0. Here, the CALL routine is placed in a DO loop, which iterates until the routine returns a value of 0 for the length. The number of words is, therefore, one fewer than this number. Here is the listing:

```
Listing of Data Set COUNT

                    Num_
String              words

ONE TWO THREE         3
ONE TWO               2
ONE                   1
```

Functions That Substitute Letters or Words in Strings

TRANSLATE can substitute one character for another in a string. TRANWRD can substitute a word or several words for one or more words. TRANWRD is especially useful in such tasks as address standardization.

Function: TRANSLATE

Purpose   To exchange one character value for another. For example, you might want to change values 1–5 to the values A–E.

Syntax: TRANSLATE(*character-value, to-1, from-1 <,… to-n, from-n>*)

character-value is any SAS character value.

to-n is a single character or a list of character values.

from-n is a single character or a list of characters.

Each character listed in *from-n* is changed to the corresponding value in *to-n*. If a character value is not listed in *from-n*, it will be unaffected. (Most SAS programmers consider the order of the second two arguments (*from* and *to*) to be in reverse order.)

Examples

In these examples, CHAR = "12X45", ANS = "Y".

Function	Returns
TRANSLATE(CHAR,"ABCDE","12345")	"ABXDE"
TRANSLATE(CHAR,'A','1','B','2','C','3','D','4','E','5')	"ABXDE"
TRANSLATE(ANS,"10","YN")	"1"

Program 1.47: Converting values of '1','2','3','4', and '5' to 'A','B','C','D', and 'E' respectively

```
***Primary function: TRANSLATE;

data multiple;
   input Ques : $1. @@;
   Ques = translate(Ques,'ABCDE','12345');
datalines;
1 4 3 2 5
5 3 4 2 1
;
title "Listing of Data Set MULTIPLE";
proc print data=multiple noobs;
run;
```

Explanation

In this example, you want to convert the character values of 1–5 to the letters A–E. The two arguments in this function seem backwards to this author. You would expect the order to be "from–to" rather than the other way around. I suppose others at SAS felt the same way, since a more recent function, TRANWRD (next example), uses the "from–to" order for its arguments. While you could use a format along with a PUT function to do this translation, the TRANSLATE function is more compact and easier to use in cases like this. A listing of data set MULTIPLE follows:

```
Listing of Data Set MULTIPLE

Ques

   A
   D
   C
   B
   E
   E
   C
   D
   B
   A
```

Program 1.48: Converting the values "Y" and "N" to 1's and 0's

```
***Primary functions: TRANSLATE, UPCASE
***Other functions: INPUT;

data yes_no;
   length Char $ 1;
   input Char @@;
   x = input(
       translate(
       upcase(Char),'01','NY'),1.);
datalines;
N Y n y A B 0 1
;
title "Listing of Data Set YES_NO";
proc print data=yes_no noobs;
run;
```

Explanation

This rather silly program was written mainly to demonstrate the TRANSLATE and UPCASE functions. A couple of IF statements, combined with the UPCASE function in the DATA step, would probably be more straightforward. In this program, the UPCASE function converts lowercase values of "n" and "y" to their uppercase equivalents. The TRANSLATE function then converts the N's and Y's to the characters "0" and "1," respectively. Finally, the INPUT function does the character to numeric conversion. Note that the data values of "1" and "0" do not get translated, but do get converted to numeric values. As you can see in the following listing, the program does get the job done.

```
Listing of Data Set YES_NO

Char    x

  N     0
  Y     1
  n     0
  y     1
  A     .
  B     .
  0     0
  1     1
```

Function: **TRANWRD**

Purpose: To substitute one or more words in a string with a replacement word or words. It works like the find-and-replace feature of most word processors.

Syntax: TRANWRD (*character-value, find, replace*)

character-value is any SAS character value.

find is one or more characters that you search for in the *character-value* and want to replace with the character or characters in the *replace* value.

replace is one or more characters that replace the entire *find* value.

Making the analogy to the find and replace feature of most word processors here, *find* represents the string you are searching for and *replace* represents the string that will replace the found value. Notice that the order of *find* and *replace* in this function is opposite from the order in the TRANSLATE function (and more logical to this author).

If the *replace* value is longer than the *find* value, the result may be truncated. Be sure that the length of the result is long enough.

Examples

For these examples, STRING = "123 Elm Road", FIND = "Road", and REPLACE = "Rd.".

Function	Returns
TRANWRD(STRING,FIND,REPLACE)	"123 Elm Rd."
TRANWRD("Now is the time","is","is not")	"Now is not the time"
TRANWRD("one two three","four","4")	"one two three"
TRANWRD("Mr. Rogers","Mr."," ")	" Rogers"
TRANWRD("ONE TWO THREE","ONE TWO","A B")	"A B THREE"

Program 1.49: Converting words such as Street to their abbreviations such as St. in an address

```
***Primary function: TRANWRD;

data convert;
   input @1 address $20. ;
   *** Convert Street, Avenue and Road to their abbreviations;
   Address = tranwrd(Address,'Street','St.');
   Address = tranwrd (Address,'Avenue','Ave.');
   Address = tranwrd (Address,'Road','Rd.');
datalines;
89 Lazy Brook Road
123 River Rd.
12 Main Street
;
title 'Listing of Data Set CONVERT';
proc print data=convert;
run;
```

Explanation

TRANWRD is an enormously useful function. This example uses it to help standardize a mailing list, substituting abbreviations for full words. Another use for this function is to make the *replace* string a blank, thus allowing you to remove words such as Jr. or Mr. from an address. The converted addresses are shown in the following listing:

```
Listing of Data Set CONVERT

Obs    address

1      89 Lazy Brook Rd.
2      123 River Rd.
3      12 Main St.
```

Functions That Compute the Length of Strings

The following four functions compute the length of character values. The LENGTH function (the oldest of the lot) does not count trailing blanks in its calculation. The LENGTHN function is identical to the LENGTH function with one exception: if there is a null string (technically speaking, a string consisting of all blanks), the LENGTH function returns a value of 1 while the LENGTHN function returns a value of 0. I would recommend using the LENGTHN function as your general purpose length function in place of the older LENGTH function. The LENGTHC function operates like the LENGTH function except it counts trailing blanks in its computation. Finally, LENGTHM computes the length used to store this variable in memory. In most applications, the LENGTHM and LENGTHC functions return the same value. You may see some differences when working with macro variables. The LENGTHC function is a useful way to determine the storage length of a character variable (instead of using PROC CONTENTS, for example).

Function: **LENGTH**

Purpose: To determine the length of a character value, not counting trailing blanks. A null argument returns a value of 1.

Syntax: LENGTH(*character-value*)

character-value is any SAS character value.

Examples

For these examples, CHAR = "ABC   ".

Function	Returns
LENGTH("ABC")	3
LENGTH(CHAR)	3
LENGTH(" ")	1

Function: **LENGTHC**

Purpose: To determine the length of a character value, including trailing blanks.

Syntax: LENGTHC (*character-value*)

character-value is any SAS character value.

Examples
For these examples, CHAR = "ABC   ".

Function	Returns
LENGTHC ("ABC")	3
LENGTHC (CHAR)	6
LENGTHC (" ")	1

Function: **LENGTHM**

Purpose: To determine the length of a character variable in memory.

Syntax: LENGTHM (*character-value*)

character-value is any SAS character value.

Examples
For these examples, CHAR = "ABC   ".

Function	Returns
LENGTHM ("ABC")	3
LENGTHM (CHAR)	6
LENGTHM (" ")	1

Function: **LENGTHN**

Purpose: To determine the length of a character value, not counting trailing blanks. A null argument returns a value of 0.

Syntax: LENGTHN (*character-value*)

character-value is any SAS character value.

Examples

For these examples, CHAR = "ABC   ".

Function	Returns
LENGTHN ("ABC")	3
LENGTHN (CHAR)	3
LENGTHN (" ")	0

Program 1.50: Demonstrating the LENGTH, LENGTHC, LENGTHM, and LENGTHN functions

```
***Primary functions: LENGTH, LENGTHC, LENGTHM, LENGTHN;

data length_func;
   file print;
   Notrail = "ABC";
   Trail   = "DEF    ";  * Three trailing blanks;
   Null    = " ";        * Null string;
   Length_notrail = length(Notrail);
   Length_trail   = length(Trail);
   Length_null    = length(Null);
   Lengthc_notrail = lengthc(Notrail);
   Lengthc_trail  = lengthc(Trail);
   Lengthc_null   = lengthc(Null);
   Lengthn_notrail = lengthn(Notrail);
   Lengthn_trail  = lengthn(Trail);
   Lengthn_null   = lengthn(Null);
   put "Notrail:" Notrail /
       "Trail:" Trail /
       "Null:" Null / 25*'-'/
       (Length_:) (=) /
       (Lengthc:) (=) /
       (Lengthn:) (=) ;
 run;
```

Explanation

The LENGTH and LENGTHN functions return the length of a character variable, **not counting trailing blanks**. The only difference between the LENGTH and LENGTHN functions is that the LENGTH function returns a value of 1 for a null string, while the LENGTHN function returns a 0. The LENGTHC function **does count trailing blanks** in its calculations. Finally, the LENGTHM function returns the number of bytes of memory used to store the variable. Notice in this program that the LENGTHM and LENGTHC functions yield the same value. Look over the following listing to be sure you understand the differences among these functions:

```
Notrail:ABC
Trail:DEF
Null:
------------------------
Length_notrail=3 Length_trail=3 Length_null=1
Lengthc_notrail=3 Lengthc_trail=6 Lengthc_null=1
Lengthn_notrail=3 Lengthn_trail=3 Lengthn_null=0
```

Functions That Count the Number of Letters or Substrings in a String

The COUNT function counts the number of times a given substring appears in a string. The COUNTC function counts the number of times specific characters occur in a string. Finally, COUNTW counts the number of "words" (i.e. characters separated by delimiters) in a character value.

Function: COUNT

Purpose: To count the number of times a given substring appears in a string. With the use of a modifier, case can be ignored. If no occurrences of the substring are found, the function returns a 0.

Syntax: COUNT(*character-value, find-string <,'modifiers'>*)

character-value is any SAS character value.

find-string is a character variable or SAS string literal to be counted. The following *modifiers*, placed in single or double quotation marks, can be used with COUNT:

```
i or I    ignore case.

t or T    ignore trailing blanks in both the character value and the
          find-string.
```

Examples

For these examples, STRING1 = "How Now Brown COW" and STRING2 = "ow".

Function	Returns
COUNT(STRING1, STRING2)	3
COUNT(STRING1,STRING2,'I')	4
COUNT(STRING1, "XX")	0
COUNT("ding and dong","g ")	1
COUNT("ding and dong","g ","T")	2

Program 1.51: Using the COUNT function to count the number of times the word "the" appears in a string

```
***Primary Function: COUNT;

data dracula;
   input String $char60.;
   Num = count(String,"the");
   num_no_case = count(String,"the",'i');
datalines;
The number of times "the" appears is the question
THE the
None on this line!
There is the map
;
title "Listing of Data Set DRACULA";
proc print data=dracula noob;
run;
```

Explanation

In this program, the COUNT function is used with and without the i (ignore case) modifier. In the first observation, the first "The" has an uppercase T, so it does not match the substring and is not counted for the variable NUM. But when the i modifier is used, it does count. The same holds for the second observation. When there are no occurrences of the substring, as in the third observation, the function returns a 0. The fourth line of data demonstrates that COUNT ignores word boundaries when searching for strings. Here is a listing of data set DRACULA:

```
Listing of Data Set DRACULA

                                                    num_no_
String                                         Num    case

The number of times "the" appears is the question  2     3
THE the                                            1     2
None on this line!                                 0     0
There is the map                                   1     2
```

Function: COUNTC

Purpose: To count the number of individual characters that appear or do not appear in a string. With the use of a modifier, case can be ignored. Another modifier allows you to count characters that do not appear in the string. If no specified characters are found, the function returns a 0.

Syntax: COUNTC(*character-value, characters <,'modifiers'>*)

character-value is any SAS character value.

characters is one or more characters to be counted. It may be a string literal (letters in quotation marks) or a character variable.

The following *modifiers*, placed in quotation marks, may be used with COUNTC:

  i or I ignore case.

  o or O process the character or characters and modifiers only once. If the COUNTC function is used in the same DATA step, the previous character and modifier values are used and the current values are ignored.

  t or T ignore trailing blanks in the *character-value* or the *characters*. Note that this modifier is especially important when looking for blanks or when you are using the v modifier.

v or V count only the characters that do **not** appear in the character-value. Remember that this count will include trailing blanks unless the t modifier is used.

Examples

For these examples, STRING1 = "How Now Brown COW" and STRING2 = "wo".

Function	Returns
COUNTC("AaBbbCDE","CBA")	3
COUNTC("AaBbbCDE","CBA",'i')	6
COUNTC(STRING1, STRING2)	6
COUNTC(STRING1,STRING2,'i')	8
COUNTC(STRING1, "XX")	0
COUNTC("ding and dong","g ")	4 (2 g's and 2 blanks)
COUNTC("ding and dong","g ","t")	2 (blanks trimmed)
COUNTC("ABCDEabcde","BCD",'vi')	4 (A, E, a, and e)

Program 1.52: Demonstrating the COUNTC function to find one or more characters or to check if characters are not present in a string

```
***Primary Function: COUNTC;

data count_char;
   input String $20.;
   Num_A = countc(String,'A');
   Num_Aa = countc(String,'a','i');
   Num_A_or_B = countc(String,'AB');
   Not_A = countc(String,'A','v');
   Not_A_trim = countc(String,'A','vt');
   Not_Aa = countc(String,'A','iv');
datalines;
UPPER A AND LOWER a
abAB
BBBbbb
;
```

```
title "Listing of Data Set COUNT_CHAR";
proc print data=count_char;
run;
```

Explanation

This program demonstrates several features of the COUNTC function. The first use of the function simply looks for the number of times the uppercase letter A appears in the string. Next, by adding the i modifier, the number of uppercase or lowercase A's is counted. Next, when you place more than one character in the list, the function returns the total number of the listed characters. The v modifier is interesting. The first time it is used, COUNTC is counting the number of characters in the string that are not uppercase A's. Notice in the following listing that this count includes the trailing blanks. However, in the next statement of the program, when the v and t modifiers are used together, the trailing blanks are not counted.

```
Listing of Data Set COUNT_CHAR


                                       Num_A_           Not_A_
Obs   String              Num_A  Num_Aa  or_B   Not_A   trim   Not_Aa

 1    UPPER A AND LOWER a    2      3      2      18      17      17
 2    abAB                   1      2      2      19       3      18
 3    BBBbbb                 0      0      3      20       6      20
```

Function: COUNTW

Purpose: To count the number of words in a character value. By using modifiers, you can alter the way that this function operates. If you do not specify delimiters, there is a set of defaults that are slightly different depending on whether you are using ASCII or EBCDIC (therefore, we recommend that you specify delimiters when you use this function.) If no occurrences of the substring are found, the function returns a 0.

Syntax: COUNTW(*character-value* <,*delim*> <,'*modifiers*'>)

character-value is any SAS character value.

delim is a list of delimiters. Note that if you use the k modifier, this list represents characters that are NOT delimiters.

If this argument is omitted, the default delimiters for ASCII environments (used on PC's and UNIX machines) are as follows:

blank . < (+ & ! $ *) ; ^ - / , % |

For EBCDIC environments (used mainly on main-frame computers), the default delimiters are as follows:

blank . < (+ | & ! $ *) ; ¬ - / , % | ¢

modifiers are one or more character values that either define classes of modifiers (such as all punctuation) or modify the way in which the function operates (see the following list). If you supply your own list of delimiters, any delimiters defined by modifiers are added to your list. If you want to supply modifiers, you must also specify a list of delimiters.

The following *modifiers*, placed in single or double quotation marks, can be used with COUNTW:

a	include alphabetic characters (uppercase or lowercase).
c	include control characters.
d	include digits.
h	include horizontal tab.
i	ignore case.
k	treat all characters NOT in your list of delimiters or defined by modifiers as delimiters.
l	include lowercase characters.
m	affects the way that multiple delimiters are treated. Without this modifier, multiple delimiters are treated as a single delimiter and delimiters at the beginning or end of a string are ignored. Using the 'm' modifier causes multiple delimiters to separate words of zero length. In addition, delimiters at the beginning or end of a string also delimit words of zero length.
o	process the list of modifiers only once (even if their values change). This may make this function run faster.
p	include all punctuation.
q	ignore delimiters inside of quotes.
s	include space characters (blanks, tabs, horizontal and vertical line feeds).

t	trim trailing blanks from *character-value* and the delimiter arguments.	
u	include uppercase characters.	
w	include printable characters.	
x	include hexadecimal digits.	

Examples

For these examples,
```
STRING1 = "How Now Brown COW"
STRING2 = "one,two,three,four"
STRING3 = "one,,two,,three"
```

Function	Returns
COUNTW(STRING1)	4
COUNTW(STRING1,' ')	4
COUNTW(STRING2)	4
COUNTW(STRING2,' ')	1
COUNTW(STRING3)	3
COUNTW(STRING3,' ,','m')	5

Program 1.53: Demonstrating the COUNTW function

```
***Primary function: COUNTW;
data count_words;
   input String $40.;
   Words = countw(String,,'sp');
datalines;
One two three four
One, and 3,4,5
oneword
;
title "Listing of data set COUNT_WORDS";
proc print data=count_words noobs;
run;
```

Explanation

In this program, the modifiers s (space characters) and p (punctuation) are used to select word delimiters. Remember that you need the two commas between the first argument and the modifiers. With only one comma, the function would treat the letters s and p as delimiters. Here is the output:

```
Listing of data set COUNT_WORDS

String              Words

One two three four     4
One, and 3,4,5         5
oneword                1
```

Miscellaneous String Functions

Don't be put off by the "miscellaneous" in this heading. Many of these functions are extremely useful; they just didn't fit neatly into categories.

Function: CHOOSEC

Purpose: To select a character value from a list of character values (supplied as the 2nd to *n*th arguments of the function)

Syntax: CHOOSEC(*item-number, value-1<, value-2, …>*)

item-number selects which item to return from the list of values. If the value is negative, counting starts at the end of the list.

value-n are character values from which to choose.

If the length of the resulting character value has not been defined, it will be assigned a length of 200.

Examples:

For these examples, Name1 = "Fred", Name2 = "Harriet", Name3 = "Joyce".

Function	Returns
CHOOSEC(1,Name1,Name2,Name3)	Fred
CHOOSEC(2,Name1,Name2,Name3)	Harriet
CHOOSEC(4,Name1,Name2,Name3)	missing value (_error_ = 1)
CHOOSEC(-1,Name1,Name2,Name3)	Joyce

Program 1.54: Using the CHOOSEC function to randomly select one of three reviewers for each candidate (numbered 1 to n)

```
***Primary function: CHOOSEC;
***Other functions: CEIL, RANUNI;

data reviewer;
   do Candidate = 1 to 10;
      Name = choosec(ceil(ranuni(0)*3),"Fred","Mary","Xi");
      output;
   end;
run;

title "Listing of data set REVIEWER";
proc print data=reviewer noobs;
run;
```

Explanation

First of all, my thanks to Mike Zdeb for suggesting this example and the example demonstrating CHOOSEN (to follow).

In this example, you want to assign one of three interviewers, at random, to see each candidate as they arrive. The first argument of the CHOOSEC function is a numeric value that specifies which character value to choose in the list of arguments that follow. One way to generate a random integer from 1 to n is to first use the RANUNI function to generate a random value between 0 and 1. Next you multiply this value by n to generate a random number between 0 and n. Finally, the CEIL function rounds up positive numbers to the next highest integer. For example, if the random function generates a value of .1, 3 times .1 = .3. This, rounded to the next highest integer, gives you a 1. An alternative expression for a random integer from 1 to n is (int(ranuni(0)*3 + 1). You can read more about the use of random functions in Chapter 9, "Random Number Functions."

Once you have generated a random integer from 1 to 3, the CHOOSEC function will return the name corresponding with that selection.

Here is the listing:

```
Listing of data set REVIEWER

Candidate     Name

     1        Xi
     2        Xi
     3        Fred
     4        Mary
     5        Fred
     6        Fred
     7        Mary
     8        Mary
     9        Fred
    10        Fred
```

Function: **CHOOSEN**

Purpose: To select a numeric value from a list of numeric values (supplied as the 2nd to *n*th arguments of the function).

Note: This is not really a character function, but it made some sense to include it here following the CHOOSEC function.

Syntax: CHOOSEN(*item-number, value-1<, value-2, …>*)

item-number selects which item to return from the list of values. If the value is negative, counting starts at the end of the list.

value-n are numeric values from which to choose.

Examples:

For these examples, $x = 4$, $y=7$, $z=-9$.

Function	Returns
CHOOSEN(1,x,y,z)	4
CHOOSEN(2,x,y,z)	7
CHOOSEN(4,Name1,Name2,Name3)	missing value (_error_ = 1)
CHOOSEN(-1,x,y,z)	-9

Program 1.55: Using CHOOSEN to randomly select a reviewer (by number) for each of n candidates

```
***Primary function: CHOOSEN;
***Other functions: CEIL, RANUNI;
data reviewer;
   do Candidate = 1 to 10;
      NumberN= choosen(ceil(ranuni(0)*3),850,620,103);
      output;
   end;
run;

title "Listing of data set REVIEWER";
proc print data=reviewer noobs;
run;
```

Explanation

This program is nearly identical to the previous one, except that you want to choose one of three numbers at random instead of randomly selected names. Since you want to return a numeric value, you need to use the CHOOSEN (N is for numeric) function.

Here is the listing:

```
Listing of data set REVIEWER

            Number
Candidate     N

    1        850
    2        620
    3        103
    4        103
    5        850
    6        850
    7        620
    8        103
    9        103
   10        850
```

Function: **MISSING**

Purpose: To determine if the argument is a missing (character or numeric) value. This is a handy function to use, since you don't have to know if the variable you are testing is character or numeric. The function returns a 1 (true) if the value is a missing value, a 0 (false) otherwise.

Syntax: **MISSING(*variable*)**

variable is a character or numeric variable or expression.

Examples:

For these examples, NUM1 = 5, NUM2 = ., CHAR1 = "ABC", and CHAR2 = " ".

Function	Returns
MISSING(NUM1)	0
MISSING(NUM2)	1
MISSING(CHAR1)	0
MISSING(CHAR2)	1

Program 1.56: Determining if there are any missing values for all variables in a data set

```
***Primary function: MISSING
***Other function: DIM;

***First, create a data set for testing;
data test_miss;
   input @1 (X Y Z)(1.)
         @4 (A B C D)($1.);
datalines;
123ABCD
..7 FFF
987RONC
;
title "Count of Missing Values";
data find_miss;
   set test_miss end=Last;
   array nums[*] _numeric_;
   array chars[*] _character_;
   do i = 1 to dim(nums);
      if missing(nums[i]) then NN + 1;
   end;
   do i = 1 to dim(chars);
      if missing(chars[i]) then NC + 1;
   end;
   file print;
   if Last then put NN "numeric and " NC "character values missing";
run;
```

Explanation

Notice that the MISSING function can take either a numeric or a character argument. In this program, since you need to have separate arrays for the character and numeric variables, you could have just as easily used the standard period and blank to represent missing values. Because of the END= option in the SET statement, the program outputs the counts when the last observation is processed from the data set TEST_MISS. Here is the output from this program:

```
Count of Missing Values

2 Numeric and 1 Character values missing
```

One additional feature of the MISSING function is that it returns a value of "true" for all numeric missing values (that is, the additional 27 numeric missing values .A, .B, … .Z, and ._). Since some programmers may be unfamiliar with these alternative missing values, I have included another example that uses values of .A and .B to keep track of two different type of missing values from a survey (Not Applicable; Did not answer). Here is the program:

Program 1.57: Demonstrating the MISSING function with .A, .B, etc. numeric values

```
***Primary function: MISSING;
*Create a data set with alternate missing values:
 Code of 888 means "Not applicable"
 Code of 999 means "Did not answer";
proc format;
   invalue readit 888 = .A
                  999 = .B
               other  = _same_ ;
   value miss .A = "Not applicable"
              .B = "Did not answer";
run;
data alternate_missing;
   input Value : readit. @@;
   format Value miss.;
   *Count total number of missing values;
   if missing(Value) then Miss_count + 1;
datalines;
100 200 . 888 300 999 999 600
;
title "Listing of data set ALTERNATE_MISSING";
proc print data=alternate_missing noobs;
run;
```

Explanation

As you can see from the user-defined informat, a code of 888 was used to indicate that the question was "Not applicable" while a code of 999 was used to indicate that the respondent "Did not answer." In the DATA step, the MISSING function increments the count of missing values, even when these values are equal to .A or .B. Here is the output from this program:

```
Listing of data set ALTERNATE_MISSING

                 Miss_
    Value        count

          100      0
          200      0
            .      1
Not applicable     2
          300      2
Did not answer     3
Did not answer     4
          600      4
```

Function: COALESCEC

Purpose: To select the first non-missing character value from a list of character values.

If not previously defined, the length of the returned variable will be 200.

Syntax: COALESCEC (*<of> Char-1 <,Char-n>*)

Char-1, Char-n is a list of character values. The function returns the first non-missing value in the list. If all the values are missing (have a zero length or consist of all blanks), the function returns a missing value. If a list in the form VAR1-VARN is used, you must precede the list with the word "of".

Examples

For these examples, Name1=" ", Name2="Hugo", Name3="Elvis".

Function	Returns
COALESCEC (Name1,Name2,Name3)	"Hugo"
COALESCEC (" ","B","C")	"B"
COALESCEC (" "," "," ")	missing value
COALESCEC (of Name1-Name3)	"Hugo"

Program 1.58: Demonstrating the COALESCEC function

```
***Primary function: COALESCEC;
data first_nonmissing;
    length First_nonmiss $ 10;
    input (Name1-Name3) (: $10.);
    First_nonmiss = coalescec(of Name1-Name3);
datalines;
Able Baker Charlie
. . Martin
Ron . Roger
;

title "Listing of data set FIRST_MISSING";
proc print data=first_nonmissing;
run;
```

Explanation

You want to return the first non-missing name in a list of names. Note that you need a LENGTH statement to define the length of the returned value. Also, since you are using Name1–Name3, you need to place the "of" in front of the list. Here is the listing:

```
Listing of data set FIRST_MISSING

       First_
Obs    nonmiss    Name1    Name2    Name3

 1     Able       Able     Baker    Charlie
 2     Martin                       Martin
 3     Ron        Ron               Roger
```

Function: IFC

Purpose: To select a character value based on the evaluation of a logical expression. The first argument to the function is a logical expression that is either true, false, or missing. The function returns the value of the second argument when the expression is true, the value of the third argument when the expression is false, and the value of the fourth argument if the expression is missing.

If not previously defined, the length of the returned variable will be 200.

Syntax: IFC(*logical-expression, when-true, when-false <,when-missing>*)

logical-expression is a numeric value.

when-true is a character value that is returned by the function when the *logical-expression* is true.

when-false is a character value that is returned by the function when the *logical-expression* is false.

when-missing is a character value that is returned by the function when the *logical-expression* is a missing value.

Examples

For these examples, SCORE1 = 95, SCORE2 = 60, N1 = "Great", N2 = "Not Great".

Function	Returns
IFC(SCORE1>85,N1,N2)	"Great"
IFC(SCORE2>85,N1,N2)	"Not Great"
IFC(1,"ABC","DEF")	"ABC"
IFC(.,"ABC","DEF","GHI")	"GHI"

Program 1.59: Using the IFC function to select a character value, based on the value of a logical expression

```
***Primary function: IFC;

data  testscore;
   length Category $ 9;
   input Name : $10. Score;
   if not missing(Score) then
      Category = ifc(Score ge 90, 'Excellent','Below 90');
datalines;
Ron 99
Mike 78
Susan .
```

```
Pete 91
George 65
;

title "Listing of data set TESTSCORE";
proc print data=testscore noobs;
run;
```

Explanation

For all observations when SCORE is greater than or equal to 90, the function returns a value of "Excellent." All other values of SCORE make the logical expression false so that the value "Below 90" is returned. Note that a missing value for SCORE does not result in the expression being missing.

Here is the output from the program:

```
Listing of data set TESTSCORE

Name      Score    Category

Ron          99    Excellent
Mike         78    Below 90
Susan         .
Pete         91    Excellent
George       65    Below 90
```

Note: Mike Zdeb, one of my premier reviewers, suggested a modification of the program to produce a different value of CATEGORY when SCORE is missing. Take a look at the following code:

```
data  testscore;
   length Category $ 9;
   input Name : $10. Score;
   Category = ifc((Score ge 90)*Score,'Excellent','Below 90','No Score');
datalines;
```

Why does this work? The expression SCORE GE 90 is false when SCORE is missing. In order to make the logical expression missing when SCORE is missing, you can multiply the expression by SCORE. This way, when SCORE is missing, the logical expression is also missing (and the function selects the fourth argument). What happens when SCORE is not missing? If it is below 90, SCORE GE 90 is false (0) and zero times SCORE is still false. If SCORE is greater than or equal to 90, SCORE GE 90 is true (1) and 1 times SCORE is true since all numeric values that are not missing or equal to 0 are true.

Function: RANK

Purpose: To obtain the relative position of the ASCII (or EBCDIC) characters. This can be useful if you want to associate each character with a number so that an ARRAY subscript can point to a specific character.

Syntax: RANK(*letter*)

letter can be a string literal or a SAS character variable. If the literal or variable contains more than one character, the RANK function returns the collating sequence of the first character in the string.

Examples

For these examples, STRING1 = "A" and STRING2 = "XYZ".

Function	Returns
RANK(STRING1)	65
RANK(STRING2)	88
RANK("X")	88
RANK("a")	97

Program 1.60: Using the collating sequence to convert plain text to Morse Code

```
***Primary function: RANK
***Other functions: LENGTHN, UPCASE, SUBSTR;

title "Morse Code Conversion Using the RANK Function";
data _null_;
  array Dot_dash[26] $ 4 _temporary_
    ('.-' '-...' '-.-.' '-..' '.'
     '..-.' '--.' '....' '..' '.---'
     '-.-' '.-..' '--' '-.' '---' '.--.'
     '--.-' '.-.' '...' '-' '..-'
     '...-' '.--' '-..-' '-.--' '--..');
  input @1 String $80.;
  file print;
  do i = 1 to lengthn(String);
    Letter = upcase(substr(String,i,1));
    if missing(Letter) then put letter @;
    else  do;
      Num = rank(Letter) - 64;
```

```
        put Dot_dash[num] ' ' @;
      end;
    end;
    put;
datalines;
This is a test SOS
Now is the time for all good men
;
```

Explanation

The RANK function returns a value of 65 for an uppercase A, a value of 66 for a B, and so forth (in the ASCII character set). If you subtract 64 from the RANK value of the letters A to Z, you will get the numbers 1 to 26. Each element in the temporary array is the Morse Code equivalent of the 26 letters of the alphabet.

The DO loop starts from 1 to the LENGTH of STRING. Each letter is converted to uppercase and its order in the alphabet is returned by the expression RANK(LETTER) – 64. This value is then used as the subscript in the DOT_DASH array and the appropriate series of dots and dashes is written to the output screen. As an "exercise for the reader," this problem can also be solved in an elegant manner using a user-defined format mapping the letters of the alphabet to the Morse equivalents. Here is the output from this program:

```
Morse Code Conversion Using the RANK Function
-   ....  ..  ...     ..  ...     .-    -  .  ...   -     ...  ---  ...
-.  ---  .--     ..  ...     -   ....  .    -  ..  --   .    ..-.  ---  .-.
    .-   .-..  .-..     --.  ---  ---  -..     --   .   -.
```

Function: **REPEAT**

Purpose: To make multiple copies of a string.

Syntax: **REPEAT(*character-value*, *n*)**

character-value is any SAS character value.

n is the number of repetitions. The result of this function is the original string plus *n* repetitions. Thus if *n* equals 1, the result will be two copies of the original string in the result. If you do not declare the length of the character variable holding the result of the REPEAT function, it will default to 200.

Examples

For these examples, STRING = "ABC".

Function	Returns
REPEAT(STRING,1)	"ABCABC"
REPEAT("HELLO ",3)	"HELLO HELLO HELLO HELLO"
REPEAT("*",5)	"******"

Program 1.61: Using the REPEAT function to underline output values

```
***Featured Function: REPEAT;

title "Demonstrating the REPEAT Function";
data _null_;
   file print;
   length Dash $ 50;
   input String $50.;
   if _n_ = 1 then put 50*"*";
   Dash = repeat("-",length(String) - 1);
   put String / Dash;
datalines;
Short line
This is a longer line
Bye
;
```

Explanation

I must admit, I had a hard time coming up with a reasonable program to demonstrate the REPEAT function. The preceding program underlines each string with the same number of dashes as there are characters in the string. Since you want the line of dashes to be the same length as the string, you subtract one from the length, remembering that the REPEAT function results in *n* + 1 copies of the original string (the original plus *n* repetitions).

When using the REPEAT function, remember to always make sure you have defined a length for the resulting character variable, and remember that the result of the REPEAT function is $n + 1$ repetitions of the original string. Here is the output from the program:

```
Demonstrating the REPEAT Function
**************************************************
Short line
----------
This is a longer line
--------------------
Bye
---
```

Function: **REVERSE**

Purpose: To reverse the order of text of a character value.

Syntax: REVERSE(*character-value*)

character-value is any SAS character value.

Examples
For these examples, STRING1 = "ABCDE" and STRING2 = "XYZ   ".

Function	Returns
REVERSE(STRING1)	"EDCBA"
REVERSE(STRING2)	" ZYX"
REVERSE("1234")	"4321"

Program 1.62: Using the REVERSE function to create backwards writing

```
***Primary function: REVERSE;

data backwards;
   input @1 String $char10.;
   Gnirts = reverse(String);
datalines;
Ron Cody
   XYZ
```

```
ABCDEFG
        x
1234567890
;
title "Listing of Data Set BACKWARDS";
proc print data=backwards noobs;
run;
```

Explanation

It is important to realize that if you don't specify the length of the result, it will be the same length as the argument of the REVERSE function. Also, if there were trailing blanks in the original string, there will be leading blanks in the reversed string. Look specifically at the last two observations in the following listing to see that this is the case.

```
Listing of Data Set BACKWARDS

  String        Gnirts

Ron Cody        ydoC noR
   XYZ             ZYX
ABCDEFG         GFEDCBA
        x       x
1234567890    0987654321
```

Function: **NLITERAL**

Purpose: Converts a string to a name literal if the string is not a valid V7 variable name.

Valid SAS variable names with the system option VALIDVARNAME set to V7 follow these rules:

- They must begin with a letter or underscore.
- The remaining characters are letters, numbers, or underscore.
- Length must not exceed 32 characters.
- Reserved names such as _N_, _numeric_, _character_, _all_, and _error_ are not allowed.

A name literal can be used as a variable name if the system option VALIDVARNAME is set to ANY. Name literals are placed in single or double quotation marks, followed by an uppercase or lowercase N. They can contain blanks and special characters, but cannot exceed a length of 32 characters.

Syntax: NLITERAL(*character-value*)

character-value is any SAS character value.

The NLITERAL function will place the *character-value* in single quotes if it contains an & or a % sign. Also, if the *character-value* contains more double quotes than single quotes, it will be placed in single quotes. If neither of these conditions is met, the result is placed in double quotes.

Examples

For these examples, STRING1 = "Valid123" and STRING2 = 'Black&White'.

Function	Returns
NLITERAL(STRING1)	Valid123
NLITERAL(STRING2)	'Black&White'N
NLITERAL("The % Sign")	'The % Sign'N

Program 1.63: Demonstrating the NLITERAL function

```
***Primary function: NLITERAL;

data literal;
   input @1 Varname $32. Value;
   SASName = nliteral(Varname);
datalines;
Valid123                     100
In valid                     200
Black&White                  300
'Single Quotes'              400
"Double Quotes"              500
Contains%                    600
;
```

```
title "Listing of data set LITERAL";
proc print data=literal noobs;
run;
```

Explanation

Notice that valid V7 variable names are left unchanged. Each of the other variable names are converted to name literals, with the decision of placing the result in single or double quotes based on the rules stated previously. Here is the listing:

```
Listing of data set LITERAL

Varname            Value    SASName

Valid123            100     Valid123
In valid            200     "In valid"N
Black&White         300     'Black&White'N
'Single Quotes'     400     "'Single Quotes'"N
"Double Quotes"     500     '"Double Quotes"'N
Contains%           600     'Contains%'N
```

Function: NVALID

Purpose: Determines whether a character value conforms to either the V7 or ANY naming convention. It returns values of true or false.

Syntax: NVALID(*character-value <,validvarname>*)

character-value is any SAS character value.

validvarname is an optional argument that can be either 'ANY', 'V7', or 'NLITERAL'. If it is set to 'ANY' and the *character-value* contains 32 or fewer characters, including blanks, it will return a value of 'true'. If it is set to 'V7', it accepts as valid only names meeting the V7 conventions (see the following paragraphs). If it is set to 'NLITERAL', only V7 names or name literals are accepted as valid.

If you do not include the *validvarname* option, the execution of the function will depend on the value of the system option VALIDVARNAME (V7, ANY, or UPPER).

Valid SAS variable names with the system option VALIDVARNAME set to V7 follow these rules:

- They must begin with a letter or underscore.
- The remaining characters are letters, numbers, or underscore.
- Length must not exceed 32 characters.
- Reserved names such as _N_, _numeric_, _character_, _all_, and _error_ are not allowed.

Examples

For these examples, STRING1 = "Valid123" and STRING2 = "Black&White". System option VALIDVARNAME is set to V7.

Function	Returns
NVALID(STRING1)	True
NVALID(STRING2)	False
NVALID("The % Sign")	False

Program 1.64: Demonstrating the NVALID function

```
***Primary function: NVALID;
options validvarname=v7;
data valid_names;
   input @1 Varname $32.;
   if nvalid(Varname) then Result_def = 'OK    ';
   else Result_def = 'Not OK';
   if nvalid(Varname, 'V7') then Result_V7 = 'OK    ';
   else Result_V7 = 'Not OK';
   if nvalid(Varname,'ANY') then Result_ANY = 'OK    ';
   else Result_ANY = 'Not OK';
   if nvalid(Varname,'Nliteral') then Result_nlit = 'OK    ';
   else Result_nlit = 'Not OK';
datalines;
Valid123
Contains blank
Black&White
;
title "Listing of data set VALID_NAMES";
title2 "Validvarnames set equal to V7";
proc print data=valid_names noobs;
run;
```

```
options validvarname=ANY;
data valid_names;
   input @1 Varname $32.;
   if nvalid(Varname) then Result_def = 'OK    '; [='
   else Result_def = 'Not OK';
   if nvalid(Varname, 'V7') then Result_V7 = 'OK    ';
   else Result_V7 = 'Not OK';
   if nvalid(Varname,'ANY') then Result_ANY = 'OK    ';
   else Result_ANY = 'Not OK';
   if nvalid(Varname,'Nliteral') then Result_nlit = 'OK    ';
   else Result_nlit = 'Not OK';
datalines;
Valid123
Contains blank
Black&White
;
title "Listing of data set VALID_NAMES";
title2 "Validvarnames set equal to ANY";
proc print data=valid_names noobs;
run;
```

Explanation

The program was first run with the system option VALIDVARNAME set to V7 and the
second time with this option set to ANY. Here is the listing:

```
Listing of data set VALID_NAMES
Validvarnames set equal to V7


                Result_     Result_     Result_     Result_
Varname           def         V7          ANY        nlit

Valid123        OK          OK          OK          OK
Contains blank  Not OK      Not OK      OK          OK
Black&White     Not OK      Not OK      OK          OK

Listing of data set VALID_NAMES
Validvarnames set equal to ANY


                Result_     Result_     Result_     Result_
Varname           def         V7          ANY        nlit

Valid123        OK          OK          OK          OK
Contains blank  OK          Not OK      OK          OK
Black&White     OK          Not OK      OK          OK
```

The next program shows how you can use the NVALID function to test the variable names from a SAS data set imported from a Microsoft Excel spreadsheet. In this example, the system option VALIDVARNAME was first set to ANY. If it had been set to V7, the Import Wizard would have converted all the variable names to valid V7 names (by replacing all invalid characters with underscores). Here is the program:

Program 1.65: Using the NVALID function to count the number of valid V7 and non-valid V7 variable names in a data set

```
***Primary function: NVALID;

*Note:The system option VALIDVARNAME was set to ANY before
 the Import Wizard imported the data;

options validvarname=any;
title "Listing of imported spreadsheet";
title2 "With VALIDVARNAME set to ANY";
proc print data=import_xls;
run;

proc contents data=import_xls out=cont_out noprint;
run;

title "Listing of data set CONT_OUT (Name only)";
proc print data=cont_out(keep=Name);
run;

title "Report on Variable Names";
data _null_;
   file print;
   set cont_out(keep=Name) end=last;
   if nvalid(Name,'v7') then Name_V7 + 1;
   else Name_any + 1;
   if last then put //
   "There are " Name_V7 "variables with V7 names" /
   "There are " Name_any "variables with non-V7 names";
run;
```

We will first show you a listing of data set IMPORT_XLS and the output data set produced by PROC CONTENTS (variable Name only).

```
Listing of imported spreadsheet
With VALIDVARNAME set to ANY

                     Black&    Blank    Ron's
Obs     Valid123     White     Blank     Var      N_123

 1        100         200       300      400       500
 2        101         102       103      104       105
```

```
Listing of data set CONT_OUT (Name only)

Obs     NAME

 1      Black&White
 2      Blank Blank
 3      N_123
 4      Ron's Var
 5      Valid123
```

And finally, the report produced by the DATA _Null_ program:

```
Report on Variable Names


There are 2 variables with V7 names
There are 3 variables with non-V7 names
```

C h a p t e r   2

Perl Regular Expressions

Introduction

Perl regular expressions were added in SAS®9. SAS regular expressions (similar to Perl regular expressions but using a different syntax to indicate text patterns) have actually been around since SAS 6.12, but many SAS users are unfamiliar with either SAS or Perl regular expressions. Both SAS regular expressions (the RX functions) and Perl regular expressions (the PRX functions) enable you to locate patterns in text strings. For example, you could write a regular expression to look for three digits, a dash, two digits, a dash, followed by four digits (the general form of a Social Security number). The syntax of both SAS and Perl regular expressions enables you to search for classes of characters (digits, letters, non-digits, etc.) as well as specific character values.

Since SAS already has such a powerful set of string functions, you may wonder why you need regular expressions. Many of the string processing tasks can be performed either with the traditional character functions or regular expressions. However, regular expressions can sometimes provide a much more compact solution to a complicated string manipulation task. Regular expressions are especially useful for reading highly unstructured data streams. For example, you may have a large text file and want to extract all the e-mail addresses. Another example would be extracting ZIP codes from an address file. Once a pattern is found, you can obtain the position of the pattern, extract a substring, or substitute a string.

The first edition of this book described both the Perl and SAS regular expressions, but for several reasons the older SAS regular expression functions (all beginning with RX) were not included in the second edition. First, more people may already be familiar with Perl regular expressions. Second, the implementation of the Perl regular expression in SAS is said to be more efficient. Finally, there is a more extensive set of Perl regular expression functions compared to the older SAS regular expressions. The syntax and usage of these two sets of functions is different and it is confusing to use both. If you are new to regular expressions, I suggest that you stick to the Perl regular expressions exclusively.

I have not attempted to provide a complete description of Perl regular expressions in this book. Hopefully, it will be a good start, but to become an expert, you will need to obtain a book on Perl or some other documentation on regular expressions.

A Brief Tutorial on Perl Regular Expressions

I have heard it said that Perl regular expressions are "write only." That means, with some practice, you can become fairly accomplished at writing regular expressions, but reading them, even the ones you wrote yourself, is quite difficult. I strongly suggest that you **comment** any regular expressions you write so that you will be able to change or correct your program in the future.

The PRXPARSE function is used to create a regular expression. Because this expression is compiled, it is usually placed in the DATA step following a statement such as IF _N_ = 1 then .... Since this statement is executed only once, you also need to retain the value returned by the PRXPARSE function. If the first argument to the PRXPARSE function is a constant, SAS will not recompile the regular expression for each iteration of the DATA step. However, it is still considered good programming practice to use the RETAIN statement with _N_ when you use the PRXPARSE function. So, to get started, let's take a look at the simplest type of regular expression, an exact text match.

Note: Each of these functions will be described in detail in the appropriate section of this chapter following this tutorial.

Program 2.1: Using a Perl regular expression to locate lines with an exact text match

---

```
title "Perl Regular Expression Tutorial - Program 1";
data _null_;

   if _n_ = 1 then Pattern_num = prxparse("/cat/");
   *exact match for the letters 'cat' anywhere in the string;
   retain Pattern_num;

   input String $30.;
   Position = prxmatch(Pattern_num,String);
   file print;
   put Pattern_num= String= Position=;
datalines;
there is a cat in this line.
does not match dog
cat in the beginning
at the end, a cat
cat
;
```

Explanation

You write your Perl regular expression as the argument of the PRXPARSE function. In this example, the regular expression is a constant so you need to place it in single or double quotation marks. You could, alternatively, assign the regular expression to a character variable and use that character variable as the argument of the PRXPARSE function.

The argument of the PRXPARSE function is a standard Perl regular expression. In this example, you are using the forward slashes (/) as delimiters for the regular expression. You can choose other characters as delimiters, as long as you start and end the expression with the same delimiter. The forward slash is most commonly used as a delimiter for regular expressions.

Each time you compile a regular expression, SAS assigns sequential numbers to the resulting expression. This pattern identifier returned by the PRXPARSE function is needed to identify a pattern when you perform searches using the other "PRX" functions such as PRXMATCH, PRXCHANGE, PRXNEXT, PRXSUBSTR, PRXPAREN, or PRXPOSN. Thus, the value of

PATTERN_NUM in this program is 1. If you executed another PRXPARSE function in the same DATA step, the function would return a 2.

In this simple example, the PRXMATCH function is used to return the position of the word "cat" in each of the strings. The two arguments in the PRXMATCH function are the pattern identifier from the PRXPARSE function and the string to be searched. You can think of the PRXMATCH function much as the FIND function; instead of looking for an exact match, you can also search for a text pattern. The PRXMATCH function returns the first position in String where "cat" is found. If there is no match, the PRXMATCH function returns a 0. Let's look at the following output.

```
Perl Regular Expression Tutorial - Program 1
Pattern_num=1 String=there is a cat in this line. Position=12
Pattern_num=1 String=does not match dog Position=0
Pattern_num=1 String=cat in the beginning Position=1
Pattern_num=1 String=at the end, a cat Position=15
Pattern_num=1 String=cat Position=1
```

Notice that the value of PATTERN_NUM is 1 in each observation, and the value of POSITION is the location of the letter "c" in "cat" in each of the strings. (If you invoked PRXPARSE a second time in this DATA step, the returned value would be 2.) In the second line of output, the value of POSITION is 0, since the word "cat" was not present in that string.

Be careful. Spaces count. For example, if you change the PRXPARSE line to read:

```
if _n_ = 1 then Pattern_num = prxparse("/ cat /");
```

then the output will be:

```
Perl Regular Expression Tutorial - Program 1
Pattern_num=1 String=there is a cat in this line. Position=11
Pattern_num=1 String=does not match dog Position=0
Pattern_num=1 String=cat in the beginning Position=0
Pattern_num=1 String=at the end, a cat Position=14
Pattern_num=1 String=cat Position=0
```

Notice that the strings in lines 3 and 5 no longer match because the regular expression has a space before and after the word "cat." (The reason there is a match in the fourth observation is that the length of STRING is 30 and there are trailing blanks after the word "cat.")

Perl regular expressions use special characters (called metacharacters) to represent classes of characters (named in honor of Will Rogers: "I never meta character I didn't like.") Before we present a table of Perl regular expression metacharacters, it is instructive to introduce a few of the more useful ones. The expression \d refers to any digit (0–9), \D to any non-digit, and \w to any word character (A–Z, a–z, 0–9, and _). The three repetition characters, *, +, and ? are particularly useful because they add quantity to a regular expression. For example, the * matches the preceding subexpression zero or more times; the + matches the previous subexpression one or more times, and the ? matches the previous expression zero or one times. So, here are a few examples using these characters:

`PRXPARSE("/\d\d\d/")`	matches any three digits in a row.
`PRXPARSE("/\d+/")`	matches one or more digits.
`PRXPARSE("/\w\w\w* /")`	matches any word with two or more characters followed by a space.
`PRXPARSE("/\w\w? +/")`	matches one or two word characters such as x, xy, or _X followed by one or more spaces.
`PRXPARSE("/(\w\w) +(\d) +/")`	matches two word characters, followed by one or more spaces, followed by a single digit, followed by one or more spaces. Note that the expression for the two word characters (\w\w) is placed in parentheses. Using the parentheses in this way creates what is called a **capture** buffer. The second set of parentheses (around the \d) creates the second capture buffer. Several of the Perl regular expression functions can make use of these capture buffers to extract and/or replace specific portions of a string. For example, the starting location of the two word characters or the single digit can be obtained using the PRXPOSN function.

Since the backslash, forward slash, parentheses, and several other characters have special meaning in a regular expression, you may wonder, how do you search a string that contains characters such as \, (, or)? You do this by preceding any of these special characters with a \ character (in Perl jargon called an escape character). So, to match a \ in a string, you code two backslashes like this: \\. To match an open parenthesis, you use \ (.

The following table describes several of the wild cards and metacharacters used with regular expressions:

Metacharacter	Description	Examples
. (period)	Matches exactly one character	r.n matches "ron", "run", and "ran"
\d	Matches a digit 0 to 9	\d\d\d matches any three-digit number
\D	Matches a non-digit	\D\D matches "xx", "ab" and "%%"
^	Matches the beginning of the string	^cat matches "cat" and "cats" but not "the cat"
$	Matches the end of a string	cat$ matches "the cat" but not "cat in the hat"
[xyz]	Matches any one of the characters in the square brackets	ca[tr] matches "cat" and "car"
[a-e]	Matches one of the letters in the range a to e	[a-e]\D+ matches "adam", "edam", and "car"
[a-eA-E]	Matches the letter a to e or A to E	[a-eA-E]\w+ matches "Adam", "edam," and "B13"
[^abcxyz]	Matches any characters except abcxyz	[^8]\d\d matches "123" and "999" but not "800"
x|y	Matches x or y	c(a|o)t matches "cat" and "cot"
\s	Matches a white space character, including a space or a tab,	\d+\s+\d+ matches one or more digits followed by one or more spaces, followed by one or more digits such as "123•••4" Note: •=space
\w	Matches any word character (upper- and lowercase letters, digits, and underscore)	\w\w\w matches any three word characters
\(	Matches the character (	\(\d\d\d\) matches three digits in parentheses such as "(123)"
\)	Matches the character)	\(\d\d\d\) matches three digits in parentheses such as "(123)"
\\	Matches the character \	a\\b matches a\b
\.	Matches the character .	Mr\. matches Mr.

This is not a complete list of Perl metacharacters, but it's enough to get you started. The SAS Product Documentation at http://support.sas.com/documentation or any book on Perl programming will provide you with more details. Examples of each of the "PRX" functions in this chapter will also help you understand how to write these expressions.

Note: The wording of arguments in this book might differ from the wording of arguments in the SAS Product Documentation.

Regular expression syntax includes repetition operators. These operators are extremely powerful in that they enable you to specify that an expression is to be repeated a given number of times. For example, \d is any digit. If you follow this expression with an asterisk (*), you now have a specification for zero or more digits. Likewise, \d+ is a specification for one or more digits. The following table describes these repetition operators:

Repetition Operators	Description	Examples
*	Matches the previous subexpression zero or more times	cat* matches "cat", "cats", "catanddog" c(at)* matches "c", "cat", and "catatat"
+	Matches the previous subexpression one or more times	\d+ matches one or more digits
?	Matches the previous subexpression zero or one times	hello? matches "hell" and "hello"
{n}	Matches the previous subexpression *n* times	\d{4} matches four digits
{n,m}	Matches the previous subexpression *n* or more times, but no more than *m*	\w{3,5} matches "abc" "abcd" and "abcde"

Function That Defines a Regular Expression

Function: **PRXPARSE**

Purpose: This function returns a pattern identifier that is used later by the other Perl regular expression functions or CALL routines. The identifier numbers start at 1, and each time you invoke this function, the identifier number is incremented by 1.

Syntax: PRXPARSE (*Perl-regular-expression*)

Perl-regular-expression is a Perl regular expression. See examples in the brief tutorial and in the sample programs in this chapter. If the *Perl-regular-expression* is a constant, the Perl regular expression is compiled only once. Successive calls to PRXPARSE will not cause a recompile, but will return the regular-expression-id for the regular expression that was already compiled. You may choose to execute the PRXPARSE function only once by using the code if _n_ = 1 then . . ., in which case you will need to RETAIN the returned value. (See the following examples.)

The forward slash (/) is the default delimiter. However, you can use any non-alphanumeric character instead of /. Matching brackets can also be used as delimiters. Look at the last few examples that follow to see how other delimiters can be used.

If you want the search to be case insensitive, you can follow the final delimiter with the letter i. For example, PRXPARSE ("/cat/i") will match Cat, CAT, or cat (see the fourth example).

Examples

Function	Matches	Does Not Match
PRXPARSE("/cat/")	"The cat is black"	"cots"
PRXPARSE("/^cat/")	"cat on the roof"	"The cat"
PRXPARSE("/cat$/")	"There is a cat"	"cat in the house"
PRXPARSE("/cat/i")	"The CaT"	"no dogs allowed"
PRXPARSE("/r[aeiou]t/")	"rat", "rot, "rut	"rt", "rxt", and "riot"
PRXPARSE("/\d\d\d /")	"345 " and "999 " (three digits followed by a space)	"1234" and "99"
PRXPARSE("/\d\d\d?/")	"123" and "12" (any two or three digits)	"1", "1AB", "1 9"
PRXPARSE("/\d\d\d+/")	"123" and "12345" (three or more digits)	"12X"
PRXPARSE("/\d\d\d*/")	"123", "12", "12345" (two or more digits)	"1" and "xyz"
PRXPARSE("/(\d|x)\d/")	"56" and "x9"	"9x" and "xx"
PRXPARSE("/[^a-e]\D/")	"fX", "9 ", "AA"	"aa", "99", "b%", "d9"
PRXPARSE("/^\/\//")	"//sysin dd *"	"the // is here"
PRXPARSE("/^\/(\/|*)/")	a "//" or "/*" in cols 1 and 2	"123 /*"
PRXPARSE("#//#")	"//"	"/*"
PRXPARSE("/\/\//")	"//" (equivalent to previous expression)	"/*"
PRXPARSE("[\d\d]")	any two digits	"ab"
PRXPARSE("<cat>")	"the cat is black"	"cots"

See examples of the PRXPARSE function in all of the examples in this chapter.

Functions That Locate Text Patterns

Function: PRXMATCH

Purpose: To locate the position in a string where a regular expression match is found. PRXMATCH returns the starting position at which the pattern is found. If this pattern is not found, the function returns a 0. (This function is similar to the FIND function, but instead of matching a string of characters, it is matching a pattern described by a regular expression.)

Syntax: PRXMATCH(*pattern-id* or *regular-expression, string*)

pattern-id is the value returned from the PRXPARSE function.

regular-expression is a Perl regular expression, placed in quotation marks (SAS 9.1 and later). Note: If you use *pattern-id*, you must use the PRXPARSE function first. If you place the regular expression (in quotation marks) directly in the PRXMATCH function, you do not use the PRXPARSE function.

string is a character variable or a string literal.

Examples

Regular Expression	String	Returns	Does Not Match (Returns 0)
/cat/	"The cat is black"	5	"cots"
/^cat/	"cat on the roof"	1	"The cat"
/cat$/	"There is a cat"	12	"cat in the house"
/cat/I	"The CaT"	5	"no dogs allowed"
/r[aeiou]t/	"rat", "rot, "rut	1	"rt", "rxt", "root"
/\d\d\d /	"345 " and "999 "	1	"1234" and "99"
/\d\d\d?/	"123" and "12"	1	"1", "1AB", "1 9"
/\d\d\d+/	"123" and "12345"	1	"12"
/\d\d\d*/	"123", "12", "12345"	1	"1" and "xyz"
/r.n/	"ron", "ronny", "r9n", "r n"	1	"rn"

(continued)

(continued)

Regular Expression	String	Returns	Does Not Match (Returns 0)
`/[1-5]\d[6-9]/`	`"299"`, `"106"`, `"337"`	1	`"666"`, `"919"`, `"11"`
`/(\d|x)\d/`	`"56"` and `"x9"`	1	`"9x"` and `"xx"`
`/[^a-e]\D/`	`"fX"`, `"9 "`, `"AA"`	1	`"aa"`, `"99"`, `"b%"`
`/^\/\//`	`"//sysin dd *"`	1	`"the // is here"`
`/^\/(\/|\*)/`	a `"//"` or `"/*"` in cols 1 and 2	1	`"123 /*"`

Examples of PRXMATCH without using PRXPARSE:

```
STRING = "The cat in the hat".
```

Function	Returns
`PRXMATCH("/cat/",STRING)`	4
`PRXMATCH("/\d\d+/","AB123")`	3

Program 2.2: Using a regular expression to search for phone numbers in a string

```
data phone;
   input String $char40.;
   if _n_ = 1 then Pattern = prxparse("/\(\d\d\d\) ?\d\d\d-\d{4}/");
   retain Pattern;
   if prxmatch(Pattern,String) then output;
   ***Regular expression will match any phone number in the form:
      (nnn)nnn-nnnn or (nnn) nnn-nnnn.;
   /*
      \(       matches a left parenthesis
      \d\d\d   matches any three digits
      (blank)? matches zero or one blank
      \d\d\d   matches any three digits
      -        matches a dash
      \d{4}    matches any four digits
   */
datalines;
One number (123)333-4444
Two here:(800)234-2222 and (908) 444-2344
None here
;
title "Listing of Data Set Phone";
proc print data=phone noobs;
run;
```

Explanation

To search for an open parenthesis, you use a \ (. The three \d's specify any three digits. The closed parenthesis is written as \). The space followed by the ? means zero or one space. This is followed by any three digits and a dash. Following the dash are any four digits. The notation \d{4} is a short way of writing \d\d\d\d. The number in the braces indicates how many times to repeat the previous subexpression. Since you execute the PRXPARSE function only once, remember to use the RETAIN statement to retain the value returned by the function.

Since the PRXMATCH function returns the first position of a match, any line containing one or more valid phone numbers will return a value greater than zero. Output from PROC PRINT is shown next:

```
Listing of Data Set Phone

String                                      Pattern

One number (123)333-4444                       1
Two here:(800)234-2222 and (908) 444-234       1
```

Program 2.3: Modifying Program 2.2 to search for toll-free phone numbers

```
***Primary functions: PRXPARSE, PRXMATCH;
data toll_free;
   if _n_ = 1 then
      re = prxparse("/\(8(00|77|87)\) ?\d\d\d-\d{4}\b/");
      ***regular expression looks for phone numbers of the form:
         (nnn)nnn-nnnn or (nnn) nnn-nnnn.  in addition the first
         digit of the area code must be an 8 and the next two
         digits must be either a 00, 77, or 87.;
   retain RE;
   input String $char80.;
   Position = prxmatch(RE,String);
   if Position then output;
datalines;
One number on this line (877)234-8765
No numbers here
One toll free, one not:(908)782-6354 and (800)876-3333 xxx
Two toll free:(800)282-3454 and (887) 858-1234
No toll free here (609)848-9999 and (908) 345-2222
;
title "Listing of Data Set TOLL_FREE";
proc print data=toll_free noobs;
run;
```

Explanation

Several things have been added to this program compared to the previous one. First, the regular expression now searches for numbers that begin with either (800), (877), or (887). This is accomplished by placing an "8" in the first position and then using the OR operator (the |) to select either 00, 77, or 87 as the next two digits. One other difference between this expression and the one used in the previous program is that the number is followed by a word boundary (a space or end-of-line: \b). Hopefully, you're starting to see the impressive power of regular expressions by now. Here is the listing of data set TOLL_FREE:

```
Listing of Data Set TOLL_FREE

re  String                                                    Position

1   One number on this line (877)234-8765                        25
1   One toll free, one not:(908)782-6354 and (800)876-3333 xxx   42
1   Two toll free:(800)282-3454 and (887) 858-1234               15
```

Program 2.4: Using PRXMATCH without PRXPARSE (entering the regular expression directly in the function)

```
***Primary function: PRXMATCH;

data match_it;
   input @1 String $20.;
   Position = prxmatch("/\d\d\d/",String);
datalines;
LINE 345 IS HERE
NONE HERE
ABC1234567
;
title "Listing of Data Set MATCH_IT";
proc print data=match_it noobs;
run;
```

Explanation

In this program, the regular expression to search for three digits is placed directly in the PRXMATCH function instead of a return code from PRXPARSE. (Note that many of the other PRX functions and CALL routines do not allow the regular expression as an argument.) The output is:

```
                        Listing of Data Set MATCH_IT

String              Position

LINE 345 IS HERE         6
NONE HERE                0
ABC1234567               4
```

The last line of this output is a good illustration of what Jason Secosky (the SAS guru on regular expressions) calls "false positives." You are looking for three digits and not specifying what comes after the three digits. So, even though there are more digits following the string "123", the pattern still matches. If you did not want this string to be selected by the regular expression, you could add a word boundary (\b) at the end of the regular expression (/\d\d\d\b/). A word boundary is either a space or the beginning or end of a line. If you did this, the value of POSITION in the last observation would be 0.

Function: CALL PRXSUBSTR

Purpose: Used with the PRXPARSE function to locate the starting position and length of a pattern within a string. The PRXSUBSTR CALL routine serves much the same purpose as the PRXMATCH function, plus it returns the length of the match as well as the starting position.

Syntax: CALL PRXSUBSTR(*pattern-id*, *string*, *start* <,*length*>)

pattern-id is the return code from the PRXPARSE function.

string is the string to be searched.

start is the name of the variable that is assigned the starting position of the pattern.

length is the name of a variable, if specified, that is assigned the length of the substring. If no substring is found, the value of length is zero.

Note: It is especially useful to use the SUBSTRN function (instead of the SUBSTR function) following the call to PRXSUBSTR, since using a zero length as an argument of the SUBSTRN function results in a character missing value instead of an error.

Examples

For these examples, RE = PRXPARSE("/\d+/").

Function	String	Start	Length
CALL PRXSUBSTR(RE,STRING,START,LENGTH)	"ABC 1234 XYZ"	5	4
CALL PRXSUBSTR(RE,STRING,START,LENGTH)	"NO NUMBERS HERE"	0	0
CALL PRXSUBSTR(RE,STRING,START,LENGTH)	"123 456 789"	1	3

Program 2.5: Locating all 5- or 9-digit ZIP codes in a list of addresses

Here is an interesting problem that shows the power of regular expressions. You have a mailing list. Some addresses have three lines, some have four, and others have possibly more or less. Some of the addresses have ZIP codes, some do not. The ZIP codes are either five digits or nine digits, a dash, followed by four digits (ZIP + 4). Go through the file and extract all the valid ZIP codes.

```
***Primary functions: PRXPARSE and CALL PRXSUBSTR
***Other function: SUBSTRN;

data zipcode;
   if _n_ = 1 then RE = prxparse("/ \d{5}(-\d{4})?/");
   retain RE;
   /*
      Match a blank followed by 5 digits followed by
      either nothing or a dash and 4 digits

      \d{5}     matches 5 digits
      -         matches a dash
      \d{4}     matches 4 digits
      ?         matches zero or one of the preceding subexpression

   */

   input String $80.;
   length Zip_code $ 10;
   call prxsubstr(RE,String,Start,Length);
   if Start then do;
      Zip_code = substrn(String,Start + 1,Length - 1);
```

```
      output;
   end;
   keep Zip_code;
datalines;
John Smith
12 Broad Street
Flemington, NJ 08822
Philip Judson
Apt #1, Building 7
777 Route 730
Kerrville, TX 78028
Dr. Roger Alan
44 Commonwealth Ave.
Boston, MA 02116-7364
;
title "Listing of Data Set ZIPCODE";
proc print data=zipcode noobs;
run;
```

Explanation

The regular expression is looking for a blank, followed by five digits, followed by zero or one occurrences of a dash and four digits. The PRXSUBSTR CALL routine checks if this pattern is found in the string. If it is found (START greater than zero), the SUBSTRN function extracts the ZIP code from the string. Note that the starting position in this function is START + 1 since the pattern starts with a blank. Also, since the SUBSTRN function is conditionally executed only when START is greater than zero, the SUBSTR function would be equivalent. Here is a listing of the ZIP codes:

```
Listing of Data Set ZIPCODE

Zip_code

08822
78028
02116-7364
```

Program 2.6: Extracting a phone number from a text string

```
***Primary functions: PRXPARSE, CALL PRXSUBSTR
***Other functions: SUBSTR, COMPRESS;

data extract;
    if _n_ = 1 then
        Pattern = prxparse("/\(\d\d\d\) ?\d\d\d-\d{4}/");
    retain Pattern;

    length Number $ 15;
    input String $char80.;
    call prxsubstr(Pattern,String,Start,Length);
        if Start then do;
        Number = substr(String,Start,Length);
        Number = compress(Number," ");
        output;
    end;
    keep number;
datalines;
THIS LINE DOES NOT HAVE ANY PHONE NUMBERS ON IT
THIS LINE DOES: (123)345-4567 LA DI LA DI LA
ALSO VALID (123) 999-9999
TWO NUMBERS HERE (333)444-5555 AND (800)123-4567
;
title "Extracted Phone Numbers";
proc print data=extract noobs;
run;
```

Explanation

This program is similar to Program 2.2. You use the same regular expression to test if a phone number has been found, using the PRXMATCH function. The call to PRXSUBSTR gives you the starting position of the phone number and its length (remember, the length can vary because of the possibility of a space following the area code). The values obtained from the PRXSUBSTR function are then used in the SUBSTR function to extract the actual number from the text string. Finally, the COMPRESS function removes any blanks from the string. See the following listing for the results (notice that this program extracts only the first phone number on a line if there is more than one number on a line).

```
                    Extracted Phone Numbers

  Number

(123)345-4567
(123)999-9999
(333)444-5555
```

Function: CALL PRXPOSN

Purpose: To return the position and length for a capture buffer (a subexpression defined in the regular expression). Used in conjunction with the PRXPARSE and one of the PRX search functions (such as PRXMATCH).

Syntax: CALL PRXPOSN(*pattern-id, capture-buffer-number, start <,length>*)

pattern-id is the return value from the PRXPARSE function.

capture-buffer-number is a number indicating which capture buffer is to be evaluated.

start is the name of the variable that is assigned the value of the first position in the string where the pattern from the *n*th capture buffer is found.

length is the name of the variable, if specified, that is assigned the length of the found pattern.

Before we get to the examples and sample programs, let's spend a moment discussing capture buffers. When you write a Perl regular expression, you can make pattern groupings using parentheses. For example, the pattern: /(\d+) *([a-zA-Z]+)/ has two capture buffers. The first matches one or more digits; the second matches one or more upper- or lowercase letters. These two patterns are separated by zero or more blanks.

Example

For these examples, the following lines of SAS code were submitted:

```
PATTERN = PRXPARSE("/(\d+) *([a-zA-Z]+)/");
MATCH = PRXMATCH(PATTERN,STRING);
```

Function	String	Start	Length
CALL PRXPOSN(PATTERN,1,START,LENGTH)	"abc123 xyz4567"	4	3
CALL PRXPOSN(PATTERN,2,START,LENGTH)	"abc123 xyz4567"	8	3
CALL PRXPOSN(PATTERN,1,START,LENGTH)	"XXXYYYZZZ"	0	0

Program 2.7: Using the PRXPOSN CALL routine to extract the area code and exchange from a phone number

```
***Primary functions: PRXPARSE, PRXMATCH, CALL PRXPOSN
***Other function: SUBSTR;


data pieces;
   if _n_ then RE = prxparse("/\((\d\d\d)\) ?(\d\d\d)-\d{4}/");
   /*
      \(        matches an open parenthesis
      \d\d\d    matches three digits
      \)        matches a closed parenthesis
      Blank?    matches zero or more blanks
      \d\d\d    matches three digits
      -         matches a dash
      \d{4}     matches four digits
   */
   retain RE;

   input Number $char80.;
   Match = prxmatch(RE,Number);
   if Match then do;
      call prxposn(RE,1,Area_start);
      call prxposn(RE,2,Ex_start,Ex_length);
      Area_code = substr(Number,Area_start,3);
      Exchange = substr(Number,Ex_start,Ex_length);
   end;
   drop RE;
datalines;
THIS LINE DOES NOT HAVE ANY PHONE NUMBERS ON IT
THIS LINE DOES: (123)345-4567 LA DI LA DI LA
ALSO VALID (609) 999-9999
TWO NUMBERS HERE (333)444-5555 AND (800)123-4567
;
title "Listing of Data Set PIECES";
proc print data=pieces noobs heading=h;
run;
```

Explanation

The regular expression in this program looks similar to the one in Program 2.2. Here we have added a set of parentheses around the expression that identifies the area code, \d\d\d, and the expression that identifies the exchange, \d\d\d. The PRXMATCH function returns the starting position if the match is successful, 0 otherwise. The first call to PRXPOSN requests the position of the start of the first capture buffer. Since we know that the length is 3, we do not need to include a length argument in the calling sequence. The next function call asks for the starting position and length (although we don't need the length) of the exchange (the first three digits following the area code or the blank after the area code). Note that we don't need the length of the exchange either, but it was included to illustrate how lengths are obtained. The SUBSTR function then extracts the substrings. Here is a listing of the resulting data set:

```
Listing of Data Set PIECES


Number

THIS LINE DOES NOT HAVE ANY PHONE NUMBERS ON IT
THIS LINE DOES: (123)345-4567 LA DI LA DI LA
ALSO VALID (609) 999-9999
TWO NUMBERS HERE (333)444-5555 AND (800)123-4567

          Area_                            Area_
Match     start    Ex_start   Ex_length    code     Exchange

   0        .         .            .
  17        18        22           3         123      345
  12        13        18           3         609      999
  18        19        23           3         333      444
```

Program 2.8: Using regular expressions to read very unstructured data

```
***Primary functions: PRSPARSE, PRXMATCH, CALL PRXPOSN
***Other functions: SUBSTR, INPUT;

***This program will read every line of data and, for any line
   that contains two or more numbers, will assign the first
   number to X and the second number to Y;

data read_num;
***Read the first number and second numbers on line;
   if _n_ = 1 then ret = prxparse("/(\d+) +\D*(\d+)/");
   /*
      \d+      matches one or more digits
      blank+   matches one or more blanks
      \D*      matches zero or more non-digits
      \d+      matches one or more digits
   */
   retain ret;
```

```
     input String $char40.;
     Pos = prxmatch(ret,String);
     if Pos then do;
        call prxposn(ret,1,Start1,Length1);
        if Start1 then X = input(substr(String,Start1,Length1),9.);
        call prxposn(ret,2,Start2,Length2);
        if Start2 then Y = input(substr(String,Start2,Length2),9.);
        output;
     end;
     keep String X Y;
  datalines;
  XXXXXXXXXXXXXXXXXX 9 XXXXXXX          123
  This line has a 6 and a 123 in it
  456 789
  None on this line
  Only one here: 77
  ;
  title "Listing of Data Set READ_NUM";
  proc print data=read_num noobs;
  run;
```

Explanation

This example shows how powerful regular expressions can be used to read very unstructured data. Here the task was to read every line of data and to locate any line with two or more numbers on it, and then to assign the first value to X and the second value to Y. (See the program example under the PRXNEXT function for a more general solution to this problem.) The INPUT function is used to perform a character-to-numeric conversion (see Chapter 10, "Special Functions," for more details on this function).

The following listing of READ_NUM shows that this program worked as desired. The variable STRING was kept in the data set so you could see the original data and the extracted numbers in the listing.

```
Listing of Data Set READ_NUM

String                                    X     Y

XXXXXXXXXXXXXXXXXX 9 XXXXXXX      123      9    123
This line has a 6 and a 123 in it         6    123
456 789                                  456   789
```

Function: CALL PRXNEXT

Purpose: To locate the *n*th occurrence of a pattern defined by the PRXPARSE function in a string. Each time you call the PRXNEXT routine, the next occurrence of the pattern will be identified.

Syntax: CALL PRXNEXT(*pattern-id, start, stop, position, length*)

pattern-id is the value returned by the PRXPARSE function.

start is the starting position to begin the search.

stop is the last position in the string for the search. If stop is set to −1, the position of the last non-blank character in the string is used.

position is the name of the variable that is assigned the starting position of the *n*th occurrence of the pattern or the first occurrence after start.

length is the name of the variable that is assigned the length of the pattern.

Examples

For these examples, the following statements were issued:

```
RE = PRXPARSE("/\d+/");
***Look for 1 or more digits;
STRING = "12 345 ab 6 cd";
START = 1;
STOP = LENGTH(STRING);
```

Function	Returns (1st call)	Returns (2nd call)	Returns (3rd call)
CALL PRXNEXT(RE, START, STOP, POS, LENGTH)	START = 3 STOP = 14 POS = 1 LENGTH = 2	START = 7 STOP = 14 POS = 4 LENGTH = 3	START = 12 STOP = 14 POS = 11 LENGTH = 1

Program 2.9: Finding digits in random positions in an input string using CALL PRXNEXT

```
***Primary functions: PRXPARSE, CALL PRXNEXT
***Other functions: LENGTHN, INPUT;

data find_num;
   if _n_ = 1 then ret = prxparse("/\d+/");
   *Look for one or more digits in a row;
   retain ret;

   input String $40.;
   Start = 1;
   Stop = lengthn(String);
   call prxnext(ret,Start,Stop,String,Position,Length);
   array x[5];
   do i = 1 to 5 while (Position gt 0);
      x[i] = input(substr(String,Position,Length),9.);
      call prxnext(ret,Start,Stop,String,Position,Length);
   end;
   keep x1-x5 String;
datalines;
THIS 45 LINE 98 HAS 3 NUMBERS
NONE HERE
12 34 78 90
;
title "Listing of Data Set FIND_NUM";
proc print data=find_num noobs;
run;
```

Explanation

The regular expression /\d+/ says to look for one or more digits in a string. The initial value of START is set to 1 and STOP to the length of the string (not counting trailing blanks). The PRXNEXT function is called, the value of START is set to the position of the blank after the first number, and the value of STOP is set to the length of the string. POSITION is the starting position of the first digit, and LENGTH is the number of digits in the number. The SUBSTR function extracts the digits and the INPUT function does the character-to-numeric conversion. This continues until no more digits are found (POSITION = 0). See the following listing to confirm that the program worked as expected.

```
Listing of Data Set FIND_NUM

String                          x1    x2    x3    x4    x5

THIS 45 LINE 98 HAS 3 NUMBERS   45    98    3     .     .
NONE HERE                       .     .     .     .     .
12 34 78 90                     12    34    78    90    .
```

Function: **PRXPAREN**

Purpose: To return a value indicating the largest capture buffer number that found a match. Use PRXPAREN when a Perl regular expression contains several alternative matches. You may want to use this function with the PRXPOSN function. This function is used in conjunction with PRXPARSE and PRXMATCH.

Syntax: `PRXPAREN(pattern-id)`

pattern-id is the value returned by the PRXPARSE function.

Examples

For this example, `RETURN = PRXPARSE("/(one)|(two)|(three)/")` and `POSITION = PRXMATCH(RETURN,STRING)`.

Function	String	Returns
PRXPAREN(RETURN)	"three two one"	3
PRXPAREN(RETURN)	"only one here:	1
PRXPAREN(RETURN)	"two one three"	2

Program 2.10: Demonstrating the PRXPAREN function

```
***Primary functions: PRXPARSE, PRXMATCH, PRXPAREN;

/* Orders are identified by a numeric type: 1= Retail 2=Catalog
3=Internet
   Use the sale description to identify the type of sale and set
OrderType */

data paren;
   if _n_ = 1 then
   Pattern  = prxparse("/(Retail|Store)|(Catalog)|(Internet|Web)/i");
   ***look for order type in Description field;
   retain Pattern;
```

```
   input Description $char30.;
   Position = prxmatch(Pattern,Description);
   if Position then OrderType = prxparen(Pattern);
datalines;
Order placed on Internet
Retail order
Store 123: Retail purchase
Spring catalog order
Order from specialty catalog
internet order
Web order
San Francisco store purchase
;
title "Listing of Data Set PAREN";
proc print data=paren noobs;
run;
```

Explanation

Thanks to Jason Secosky at SAS for this example. Here you are inspecting a description to
determine an order type. To make it more interesting, either Retail or Store orders are type 1,
Catalog orders are type 2, and Internet or Web orders are type 3. Since each of these
descriptions are placed in parentheses (without a \ in front of it) each of these descriptions
defines a capture buffer—Retail or Store = 1, Catalog = 2, and Internet or Web = 3. The
PRXPAREN function returns the value (1, 2, or 3) of the capture buffer that was matched.

Here is the output:

```
Listing of Data Set PAREN
                                                       Order
Pattern    Description                    Position     Type

   1       Order placed on Internet          17          3
   1       Retail order                       1          1
   1       Store 123: Retail purchase         1          1
   1       Spring catalog order               8          2
   1       Order from specialty catalog      22          2
   1       internet order                     1          3
   1       Web order                          1          3
   1       San Francisco store purchase      15          1
```

Function That Substitutes One String for Another

Function: CALL PRXCHANGE

Purpose: To substitute one string for another. One advantage of using PRXCHANGE over TRANWRD is that you can search for strings using wild cards. Note that you need to use the substitution (s) operator in the regular expression to specify the search and replacement expression (see the explanation following the program).

Syntax: CALL PRXCHANGE(*pattern-id* or *regular-expression*, *times*, *old-string* <, *new-string* <, *result-length* <, *truncation-value* <, *number-of-changes*>>>>);

pattern-id is the value returned from the PRXPARSE function.

regular-expression is a Perl regular expression, placed in quotation marks.

times is the number of times to search for and replace a string. A value of –1 will replace all matching patterns.

old-string is the string that you want to replace. If you do not specify *new-string*, the replacement will take place in *old-string*.

new-string, if specified, names the variable to hold the text after replacement. If *new-string* is not specified, the changes are made to *old-string*.

result-length is the name of the variable that, if specified, is assigned a value representing the length of the string after replacement. Note that trailing blanks in *old-string* are not copied to *new-string*.

truncation-value is the name of the variable that, if specified, is assigned a value of 0 or 1. If the resulting string is longer than the length of *new-string*, the value is 1; otherwise it is a 0. This value is useful to test if your string was truncated because the replacements resulted in a length longer than the original specified length.

number-of-changes is the name of the variable that, if specified, is
assigned a value representing the total number of replacements that were
made.

Program 2.11: Demonstrating the CALL PRXCHANGE function

```
***Primary functions: PRXPARSE, CALL PRXCHANGE;

data cat_and_mouse;
   input Text $char40.;
   length New_text $ 80;

   if _n_ = 1 then Match = prxparse("s/[Cc]at/Mouse/");
   *replace "Cat" or "cat" with Mouse;
   retain Match;

   call prxchange(Match,-1,Text,New_text,R_length,Trunc,N_of_changes);
   if Trunc then put "Note: New_text was truncated";
datalines;
The Cat in the hat
There are two cat cats in this line
;
title "Listing of CAT_AND_MOUSE";
proc print data=cat_and_mouse noobs;
run;
```

Explanation

The regular expression and the replacement string is specified in the PRXPARSE function,
using the substitution operator (the s before the first /). In this example, the regular
expression to be searched for is /[Cc]at/. This matches either "Cat" or "cat" anywhere in
the string. The replacement string is "Mouse." Since the length of NEW_TEXT was set to
80, even though the replacement of "cat" (or "Cat") with "Mouse" results in a longer string,
the new length does not exceed 80. Therefore, no truncation occurs. The –1 indicates that
you want to replace every occurrence of "Cat" or "cat" with "Mouse." If you did not supply a
NEW_TEXT variable, the replacement would be made to the original string. Here is the
output from PROC PRINT:

```
Listing of CAT_AND_MOUSE


Text

The Cat in the hat
There are two cat cats in this line


                                                                    N_of_
New_text                                    Match  R_length  Trunc  changes

The Mouse in the hat                          1       42       0       1
There are two Mouse Mouses in this line       1       44       0       2
```

Program 2.12: Demonstrating the use of capture buffers with PRXCHANGE

```
***Primary functions: PRXPARSE, CALL PRXCHANGE;

data capture;
   if _n_ = 1 then Return = prxparse("S/(\w+ +)(\w+)/$2 $1/");
   retain Return;

   input String $20.;
   call prxchange(Return,-1,String);
datalines;
Ron Cody
Russell Lynn
;
title "Listing of Data Set CAPTURE";
proc print data=capture noobs;
run;
```

Explanation

The regular expression specifies one or more word characters, followed by one or more
blanks (the first capture buffer), followed by one or more word characters (the second
capture buffer). In the substitute portion of the regular expression, the $1 and $2 expressions
refer to the first and second capture buffer, respectively. So, by placing the $2 before the $1,
the two words are reversed, as shown here.

```
Listing of Data Set CAPTURE

Return    String

   1      Cody Ron
   1      Lynn Russell
```

Function That Releases Memory Used by a Regular Expression

Function: **CALL PRXFREE**

Purpose: To free resources that were allocated to a Perl regular expression (usually used with a test for end-of-file). If you do not call PRXFREE, the resources used will be freed when the DATA step ends.

Syntax: `CALL PRXFREE(pattern-id)`

`pattern-id` is the value returned from the PRXPARSE function.

Examples

For this example, the statement `PATTERN = PRXPARSE("/\d/")` preceded the call to PRXFREE.

Function	Returns
`PRXFREE(PATTERN)`	Sets the value returned by PRXPARSE to missing

Program 2.13: Data cleaning example using PRXPARSE and CALL PRXFREE

```
***Primary functions: PRXPARSE, PRXMATCH, CALL PRXFREE;

data invalid;
   ***Valid ID's are 1 to 3 digits with possible leading blanks;
   infile 'c:\books\functions\idnums.dat' end=last;
   if _n_ = 1 then Valid = prxparse("/\d\d\d| \d\d|  \d/");
   /*
      \d\d\d              matches three digits
      (single blank)\d\d  matches a blank followed by two digits
      (two blanks)\d      matches two blanks followed by one digit
   */
```

```
      retain Valid;
      input @1 ID $char3.;
      Pos = prxmatch(Valid,ID);
      if Pos eq 0 then output invalid;
      if last then call prxfree(Valid);
      drop Valid;
   run;
```

Explanation

In this example, valid IDs are character values of either three digits, a blank followed by two digits, or two blanks followed by one digit. The INFILE statement option END=LAST creates the logical variable (LAST), which is true when the last record is being read from the file. So, when LAST is true, a call is made to PRXFREE to release resources used by the regular expression. Note that the $CHAR3. informat is used to read the ID variable, since you want to maintain any leading blanks.

C h a p t e r 3

Sort Functions

Introduction

Two CALL routines, CALL SORTN and CALL SORTC, can sort the values of their arguments. They are frequently used with SAS arrays to place all the values of the array elements in ascending order.

These two CALL routines were included in SAS 9.1.3 but were still considered experimental. With SAS 9.2, they are considered fully supported routines.

This author remembers writing a SAS program, many years ago, that used a bubble sort to perform the actions of the CALL SORTN routine. With the introduction of the ORDINAL function, this process became much easier. Finally, with these CALL routines, sorting values within a SAS observation became trivial.

Function That Sorts Numeric Values

Function: CALL SORTN

Purpose: To sort the values of its numeric arguments in ascending order.

Syntax: `CALL SORTN(<of> Var-1 <, … Var-n>)`

Var-n is a numeric variable. If you use a variable list in the form Var1–Varn, precede the list with the word "of".

Arguments to the CALL SORTN routine cannot be constants. You may use array references as arguments to this routine as well (see the following examples):

Examples

For these examples, `x1=8, x2=2, x3=., x4=1, a=9, b=7, c=8.`
Also, an array is defined as `array nums[4] x1-x4;`

Call Routine	Returns
`CALL SORTN(of x1-x4)`	`x1=., x2=1, x3=2, x4=8`
`CALL SORTN(of nums[*])`	`x1=., x2=1, x3=2, x4=8`
`CALL SORTN(a,b,c)`	`a=7, b=8, c=9`

Program 3.1: Program to grade quizzes, dropping the two lowest quiz scores (using CALL SORTN)

```
***Primary function: CALL SORTN;
***Other function: MEAN;
data quiz_scores;
   input Name : $15.Quiz1-Quiz10;
   call sortn(of Quiz1-Quiz10);
   Grade = mean(of Quiz3-Quiz10);
datalines;
George 78 90 90 95 88 . 100 98 95 90
Susan 100 100 100 100 100 100 . . . .
Maxwell 50 50 90 90 95 88 87 86 84 90
;
title "Listing of data set QUIZ_SCORES";
proc print data=quiz_scores noobs heading=h;
run;
```

Explanation

Before the CALL SORTN, the values of QUIZ1–QUIZ10 are the quiz scores in the order they were read from the input data. After the call, the values of QUIZ1–QUIZ10 are the quiz scores in ascending order. Therefore, to drop the lowest two quiz scores, you take the average of QUIZ3 to QUIZ10. Note that if there are more than 2 missing quiz scores, the value of GRADE is the average of the non-missing values in QUIZ3 to QUIZ10. Notice also that the CALL routine uses the keyword "of" preceding the list of variables (otherwise, SAS would subtract the QUIZ10 value from the QUIZ3 value, and the CALL routine would only have one argument. Here is the listing of the resulting data set:

Listing of data set QUIZ_SCORES					
Name	Quiz1	Quiz2	Quiz3	Quiz4	Quiz5
George	.	78	88	90	90
Susan	.	.	.	.	100
Maxwell	50	50	84	86	87

Quiz6	Quiz7	Quiz8	Quiz9	Quiz10	Grade
90	95	95	98	100	93.25
100	100	100	100	100	100.00
88	90	90	90	95	88.75

Program 3.2: Another example of CALL SORTN using array elements as the arguments

```
***Primary function: CALL SORTN;
data top_scores;
   input Name : $15. Score1-Score8;
   array Score[8];
   call sortn(of Score[*]);
datalines;
King 8 4 6 7 9 9 9 4
Weisenbaum 9 9 8 . . 7 8 8
Chien 9 8 7 6 7 8 9 9
;
title "Listing of TOP_SCORES";
proc print data=top_scores;
  id Name;
run;
```

Explanation

This example shows how to use an array reference as the argument to the CALL routine. Notice that the keyword "of" is also needed when the array reference is used. Following the call, the eight scores are in ascending order. Here is the listing:

```
Listing of TOP_SCORES

Name        Score1 Score2 Score3 Score4 Score5 Score6 Score7 Score8

King          4      4      6      7      8      9      9      9
Weisenbaum    .      .      7      8      8      8      9      9
Chien         6      7      7      8      8      9      9      9
```

Function That Sorts Character Values

Function: CALL SORTC

Purpose: To sort the values of its character arguments in ascending order.

Syntax: `CALL SORTN(<of> Char-1 <, … Char-n>)`

Char-n is a character variable. If you use a variable list in the form Var1–Varn, precede the list with the word "of".

Arguments to the CALL SORTC routine cannot be constants. You may use array references as arguments to this routine as well (see the following examples):

Examples

For these examples, `c1='bad'`, `c2='good'`, `c3=' '`, `a1='b'`, `a2='B'`, `a3='a'`, and `a4='A'`.

Also, an array is defined as `array chars[3] c1-c3;`

Call Routine	Returns
CALL SORTC(of c1-c3)	c1=' ', c2='bad', c3='good'
CALL SORTC(of chars[*])	c1=' ', c2='bad', c3='good'
CALL SORTC(of a1-a4)	a1='A', a2='B', a3='a', a4='b'

Program 3.3: Demonstrating the CALL SORTC routine

```
***Primary function: CALL SORTC;
data names;
   input (Name1-Name5)(: $12.);
   call sortc(of Name1-Name5);
datalines;
Charlie Able Hotel Golf Echo
Zemlachenko Cody Lane Lubsen Veneziale
bbb BBB aaa aaa ZZZ
;
title "Listing of data set NAMES";
proc print data=names noobs;
run;
```

Explanation

First of all, notice that the keyword "of" is needed, even though this CALL routine operates on character values. The other feature to take note of is that alphabetic sorting is case sensitive; the uppercase letters come before the lowercase letters in the collating sequence. Here is the output:

```
Listing of data set NAMES

Name1    Name2     Name3     Name4       Name5

Able     Charlie   Echo      Golf        Hotel
Cody     Lane      Lubsen    Veneziale   Zemlachenko
BBB      ZZZ       aaa       aaa         bbb
```

Chapter 4

Date and Time Functions

Introduction

Before you start working with SAS date and time functions, remember that **SAS date values** are the number of days between January 1, 1960, and a specified date. Dates after January 1, 1960, are stored as positive numbers; dates before January 1, 1960, are stored as negative numbers. **SAS time values** are the number of seconds between midnight of the current day and another time value. **SAS datetime values** are the number of seconds between midnight, January 1, 1960, and the specified date and time. Some of the more commonly used date functions extract the day of the week, the month, or the year from a SAS date value.

Other functions deal with intervals, either the number of intervals between two dates or the date after a given number of intervals have passed. You can even compute the number of working days (the default is Saturday and Sunday as non-working days) between two dates. Making this calculation even more useful is the HOLIDAY function that, given a year, returns the date for many of the major holidays.

For situations where you only have month, day, and year values but do not have a SAS date, the MDY function can create a SAS date value, given a value for the month, day, and year. Now let's get started.

Functions That Create SAS Date, Datetime, and Time Values

The first three functions in this group of functions create SAS date values, datetime values, and time values from the constituent parts (month, day, year, hour, minute, second). The DATE and TODAY functions are equivalent and they both return the current date. The DATETIME and TIME functions are used to create SAS datetime and time values, respectively.

Function: **MDY**

Purpose: To create a SAS date from the month, day, and year.

Syntax: `MDY(month, day, year)`

month is a numeric variable or constant representing the month of the year (a number from 1 to 12).

day is a numeric variable or constant representing the day of the month (a number from 1 to 31).

year is a numeric variable or constant representing the year.

Values of month, day, and time that do not define a valid date result in a missing value, and an error message is written to the SAS log.

Examples

For these examples, M = 11, D = 15, Y = 2003.

Function	Returns
MDY(M,D,Y)	16024 (15NOV2003 – formatted value)
MDY(10,21,1980)	7599 (21OCT1980 – formatted value)
MDY(1,1,1950)	–3652 (01JAN1950 – formatted value)
MDY(13,01,2003)	numeric missing value

Program 4.1: Creating a SAS date value from separate variables representing the day, month, and year of the date

```
***Primary function: MDY;

data funnydate;
   input @1 Month  2.
         @7 Year   4.
         @13 Day   2.;
   Date = mdy(Month,Day,Year);
   format Date mmddyy10.;
datalines;
05    2000  25
11    2001  02
;
title "Listing of FUNNYDATE";
proc print data=funnydate noobs;
run;
```

Explanation

Here the values for month, day, and year were not in a form where any of the standard date informats could be used. Therefore, the day, month, and year values were read into separate variables and the MDY function was used to create a SAS date. See the following listing:

```
Listing of FUNNYDATE

Month   Year    Day         Date

    5   2000     25    05/25/2000
   11   2001      2    11/02/2001
```

**Program 4.2: Program to read in dates and set the day of the month to 15
if the day is missing from the date**

```
***Primary function: MDY;
***Other functions:  SCAN, INPUT, MISSING;

data missing;
   input @1 Dummy $10.;
   Day = scan(Dummy,2,'/');
   if not missing(Day)then Date = input(Dummy,mmddyy10.);
   else Date = mdy(input(scan(Dummy,1,'/'),2.),
                15,
             input(scan(Dummy,3,'/'),4.));
   format date date9.;
datalines;
10/21/1946
1/  /2000
01/  /2002
;
title "Listing of MISSING";
proc print data=missing noobs;
run;
```

Explanation

This program reads in a date and, when the day of the month is missing, it uses the 15th of the month. If the date was already stored as a character string in a SAS data set, this approach would work well.

The entire date is first read as a character string as the variable DUMMY. Next, the SCAN function is executed with the slash character (/) as the "word" delimiter. The second word is the month. If this is not missing, the INPUT function is used to convert the character string into a SAS date.

If DAY is missing, the MDY function is used to create the SAS date, with the value of 15 representing the day of the month. The listing follows:

```
Listing of MISSING

 Dummy       Day       Date

10/21/1946    21       21OCT1946
1/  /2000              15JAN2000
01/  /2002             15JAN2002
```

Function: **DHMS**

Purpose: To create a SAS datetime value from a SAS date value and a value for the hour, minute, and second.

Syntax: DHMS(*date, hour, minute, second*)

date is a SAS date value, either a variable or a date constant.

hour is a numerical value for the hour of the day. If hour is greater than 24, the function will return the appropriate datetime value.

minute is a numerical value for the number of minutes.

second is a numerical value for the number of seconds.

Values of the date value that are invalid result in a missing value, and an error message is written to the SAS log.

Examples

For these examples, DATE = '02JAN1960'D, H = 23, M = 15, S = 30.

Function	Returns
DHMS(DATE,H,M,S)	170130 (02JAN60:23:15:30 – formatted)
DHMS('04JUN2003'd,25,12,12)	1370394732 (05JUN03:01:12:12 – formatted)
DHMS('01JAN1960'd,0,70,0)	4200 (01JAN60:01:10:00 – formatted)

See Program 4.3.

Function: HMS

Purpose: To create a SAS time value from the hour, minute, and second.

Syntax: HMS(*hour, minute, second*)

hour is the value corresponding to the number of hours.

minute is the value corresponding to the number of minutes.

second is the value corresponding to the number of seconds.

Examples

For these examples, H = 1, M = 30, S = 15.

Function	Returns
HMS(H, M, S)	5415 (1:30:15 – formatted value)
HMS(0, 0, 23)	23 (0:00:23 – formatted value)

See Program 4.3.

Function: DATE and TODAY (equivalent functions)

Purpose: To return the current date.

Syntax: DATE() or TODAY()

Note that the parentheses are needed even though these functions do not take any arguments. (What did the TODAY function say to the MEAN function? "Don't give me any arguments!")

Examples

Note: This function was run on June 4, 2003.

Function	Returns
DATE()	15860 (04JUN2003 – formatted)
TODAY()	15860 (04JUN2003 – formatted)

See Program 4.3.

Function: DATETIME

Purpose: To return the datetime value for the current date and time.

Syntax: DATETIME()

Examples

Note: This function was run at 8:10 PM on June 4, 2004.

Function	Returns
DATETIME()	1370376600 (04JUN03:20:10:00 – formatted)

See Program 4.3.

Function: TIME

Purpose: To return the time of day when the program was run.

Syntax: TIME()

Examples

Note: This function was run at 8:10 PM.

Function	Returns
TIME()	72600 (20:10:00 – formatted)

Program 4.3: Determining the date, datetime value, and time of day

```
***Primary functions: DHMS, HMS, TODAY, DATETIME, TIME, YRDIF
***Other functions: INT;

data test;
   Date = today();
   DT = datetime();
   Time = time();
   DT2 = dhms(Date,8,15,30);
   Time2 = hms(8,15,30);
   DOB = '01jan1960'd;
   Age = int(yrdif(DOB,Date,'actual'));
   format Date DOB date9. DT DT2 datetime. Time Time2 time.;
run;

title "Listing of Data Set TEST";
proc print data=test noobs;
run;
```

Explanation

This program was run in the morning of November 10, 2009, so the values for the date, datetime, and time values correspond to that date and time.

The variable DT2 is a SAS datetime value created from the current date and specified values for the hour, minute, and second. TIME2 is a SAS time value created from three values for hour, minute, and second.

Finally, the age was computed using the YRDIF function. (See details and an important note on the YRDIF function later in this chapter.) The INT function was used to compute age as of the last birthday (it throws away all digits to the right of the decimal point). Please see the following listing:

```
Listing of Data Set TEST

    Date              DT             Time              DT2

10JUL2009      10JUL09:09:09:06    9:09:06    10JUL09:08:15:30


 Time2            DOB     Age

8:15:30     01JAN1960      49
```

Creating a Data Set to Demonstrate Other Date Functions

Run Program 4.4 to create a SAS data set called DATES. A listing of this data set follows.

Program 4.4: Program to create the DATES data set

```
data dates;
   informat Date1 Date2 date9.;
   input Date1 Date2;
   format Date1 Date2 date9.;
datalines;
01JAN1960 15JAN1960
02MAR1961 18FEB1962
25DEC2000 03JAN2001
01FEB2002 31MAR2002
;
title "Listing of Data Set DATES";
proc print data=dates noobs;
run;
```

Explanation

Although this is not a function example program, one feature should be explained: Since the INPUT statement is reading list input (i.e., one or more spaces between the data values) and since you need to supply an informat so that the values will be read as SAS date values, an INFORMAT statement precedes the INPUT statement, indicating that both variables, DATE1 and DATE2, should be read with the DATE9. informat.

```
Listing of Data Set DATES

    Date1        Date2

01JAN1960    15JAN1960
02MAR1961    18FEB1962
25DEC2000    03JAN2001
01FEB2002    31MAR2002
```

Functions That Extract the Year, Month, Day, etc. from a SAS Date

This group of functions takes a SAS date value and returns parts of the date, such as the year, the month, or the day of the week. Since these functions are demonstrated in a single program, let's supply the syntax and examples.

Function: YEAR

Purpose: To extract the year from a SAS date.

Syntax: YEAR(*date*)

date is a SAS date value.

Examples

Function	Returns
YEAR('16AUG2002'd)	2002
YEAR('16AUG02'd)	2002

See Program 4.5.

Function: QTR

Purpose: To extract the quarter (January–March = 1, April–June = 2, etc.) from a SAS date.

Syntax: QTR(*date*)

   date is a SAS date value.

Examples

Function	Returns
QTR('05FEB2003'd)	1
QTR('01DEC2003'd)	4

See Program 4.5.

Function: MONTH

Purpose: To extract the month of the year from a SAS date (1 = January, 2=February, etc.).

Syntax: MONTH(*date*)

   date is a SAS date value.

Examples

Function	Returns
MONTH('16AUG2002'd)	8

See Program 4.5.

Function: WEEK

Purpose: To extract the week number of the year from a SAS date (the week-number value is a number from 0 to 53 or 1 to 53, depending on the optional modifier).

Syntax: WEEK(*<date>* <,'*modifier*'>))

date is a SAS date value. If *date* is omitted, the WEEK function returns the week number of the current date.

modifier is an optional argument that determines how the week-number value is determined. If *modifier* is omitted, the first Sunday of the year is week 1. For dates prior to this date, the WEEK function returns a 0. The various modifiers provide several different methods for computing the value returned by the WEEK function. Most users will probably want to use this function without any modifiers. For details about the modifiers, see Product Documentation in the Knowledge Base, available at http://support.sas.com/documentation.

Examples

Function	Returns
WEEK('16AUG2002'd)	32
WEEK('01JAN1960'd)	0
WEEK('03JAN1960'd)	1
WEEK('01JAN1960'd,'V')	53

See Program 4.5 for an example.

Function: **WEEKDAY**

Purpose: To extract the day of the week from a SAS date (1 = Sunday, 2=Monday, etc.).

Syntax: WEEKDAY(*date*)

date is a SAS date value.

Examples

Function	Returns
WEEKDAY('16AUG2002'd)	5 (Thursday)

Function: **DAY**

Purpose: To extract the day of the month from a SAS date, a number from 1 to 31.

Syntax: DAY(*date*)

date is a SAS date value.

Examples

Function	Returns
DAY('16AUG2002'd)	16

See Program 4.5.

Program 4.5: Demonstrating the functions YEAR, QTR, MONTH, WEEK, DAY, and WEEKDAY

```
***Primary functions: YEAR, QTR, MONTH, WEEK, DAY, and WEEKDAY;

data date_functions;
   set dates(drop=Date2);
   Year = year(Date1);
   Quarter = qtr(Date1);
   Month = month(Date1);
   Week = week(Date1);
   Day_of_month = day(Date1);
```

```
      Day_of_week = weekday(Date1);
run;
title "Listing of Data Set DATE_FUNCTIONS";
proc print data=date_functions noobs;
run;
```

Explanation

These basic date functions are straightforward. They all take a SAS date as the single argument and return the year, the quarter, the month, the week, the day of the month, or the day of the week. Remember that the WEEKDAY function returns the day of the **week**, while the DAY function returns the day of the **month** (it's easy to confuse these two functions). A listing of DATE_FUNCTIONS follows:

```
Listing of Data Set DATE_FUNCTIONS


                                               Day_of_    Day_of_
    Date1    Year    Quarter    Month    Week    month       week

01JAN1960    1960       1         1        0       1          6
02MAR1961    1961       1         3        9       2          5
25DEC2000    2000       4        12       52      25          2
01FEB2002    2002       1         2        4       1          6
```

Functions That Extract Hours, Minutes, and Seconds from SAS Datetime and Time Values

The HOUR, MINUTE, and SECOND functions work with SAS datetime or time values in much the same way as the MONTH, YEAR, and WEEKDAY functions work with SAS date values.

Function: **HOUR**

Purpose: To extract the hour from a SAS datetime or time value.

Syntax: HOUR(*time* or *dt*)

time or *dt* is a SAS time or datetime value.

Examples

For these examples, DT = '02JAN1960:5:10:15'dt, T = '5:8:10'T.

Function	Returns
HOUR(DT)	5
HOUR(T)	5
HOUR(HMS(5,8,9))	5

See Program 4.6.

Function: **MINUTE**

Purpose: To extract the minute value from a SAS datetime or time value.

Syntax: MINUTE(*time* or *dt*)

time or *dt* is a SAS time or datetime value.

Examples

For these examples, DT = '02JAN1960:5:10:15'dt, T = '5:8:10'T.

Function	Returns
MINUTE(DT)	5
MINUTE(T)	5
MINUTE(HMS(5,8,9))	5

See Program 4.6.

Function: SECOND

Purpose: To extract the second value from a SAS datetime or time value.

Syntax: SECOND(*time* or *dt*)

time or *dt* is a SAS time or datetime value.

Examples

For these examples, DT = '02JAN1960:5:10:15'dt, T = '5:8:10'T.

Function	Returns
SECOND(DT)	15
SECOND(T)	10
SECOND(HMS(5,8,9))	9

Program 4.6: Demonstrating the HOUR, MINUTE, and SECOND functions

```
***Primary functions: HOUR, MINUTE, and SECOND;

data time;
   DT = '01jan1960:5:15:30'dt;
   T = '10:05:23't;
   Hour_dt = hour(DT);
   Hour_time = hour(T);
   Minute_dt = minute(DT);
   Minute_time = minute(T);
   Second_dt = second(DT);
   Second_time = second(T);
   format DT datetime.;
run;

title "Listing of Data Set TIME";
proc print data=time noobs heading=h;
run;
```

Explanation

The variable DT is a SAS datetime value (computed as a SAS datetime constant), and T is a SAS time value (computed as a SAS time constant). The program demonstrates that the HOUR, MINUTE, and SECOND functions can take either SAS datetime or time values as arguments. The listing follows:

```
Listing of Data Set TIME

                            Hour_  Minute_  Minute_  Second_  Second_
      DT            T    Hour_dt  time    dt      time     dt      time

01JAN60:05:15:30 36323     5      10      15       5       30       23
```

Functions That Extract the Date or Time from SAS Datetime Values

The DATEPART and TIMEPART functions extract either the date or the time from a SAS datetime value (the number of seconds from January 1, 1960).

Function: DATEPART

Purpose: To compute a SAS date from a SAS datetime value.

Syntax: DATEPART(*date-time-value*)

              date-time-value is a SAS datetime value.

Examples

For these examples, DT = '02JAN1960:5:10:15'dt.

Function	Returns
DATEPART(DT)	1 (01JAN1960 – formatted)
DATEPART('4JUN2003:20:48:15'DT)	15860 (04JUN2003 – formatted)

See Program 4.7.

Function: TIMEPART

Purpose: To extract the time part of a SAS datetime value.

Syntax: TIMEPART(*date-time-value*)

Date-time-value is a SAS datetime value.

Examples

For these examples, DT = '02JAN1960:5:10:15'dt.

Function	Returns
TIMEPART(DT)	18615 (5:10:15 – formatted)
TIMEPART('4JUN2003:20:48:15'DT)	74895 (20:48:15 – formatted)

Program 4.7: Extracting the date part and time part of a SAS datetime value

```
   ***Primary functions: DATEPART and TIMEPART;

data pieces_parts;
   DT = '01jan1960:5:15:30'dt;
   Date = datepart(DT);
   Time = timepart(DT);
   format DT datetime. Time time. Date date9.;
run;

title "Listing of Data Set PIECES_PARTS";
proc print data=pieces_parts noobs;
run;
```

Explanation

The DATEPART and TIMEPART functions extract the date and the time from the datetime value, respectively. These two functions are especially useful when you import data from other sources. (In SAS 8, imported spreadsheet columns that were formatted as dates in Microsoft Excel wound up as datetime values in the SAS data set.) You can use these two functions to separate the date and time from that value. See the following listing:

```
Listing of Data Set PIECES_PARTS

        DT               Date      Time
01JAN60:05:15:30     01JAN1960    5:15:30
```

Functions That Work with Date, Datetime, and Time Intervals

Functions in this group work with date or time intervals. The INTCK function, when used with date or datetime values, can determine the number of interval boundaries crossed between two dates. When used with SAS time values, it can determine the number of hour, minute, or second boundaries between two time values.

The INTNX function, when used with SAS date or datetime values, is used to determine the date after a given number of intervals have passed. When used with SAS time values, it computes the time after a given number of time interval units have passed.

You will find an excellent description of these two functions in *SAS Language Reference: Concepts* or in the following technical note:

http://support.sas.com/techsup/technote/ts668.html

Note: The wording of arguments in this book might differ from the wording of arguments in the Product Documentation at http://support.sas.com.

Function: INTCK

Purpose: To return the number of intervals between two dates, two times, or two datetime values. To be more accurate, the INTCK function counts the number of times a boundary has been crossed going from the first value to the second.

For example, if the interval is YEAR and the starting date is January 1, 2002, and the ending date is December 31, 2002, the function returns a 0. The reason for this is that the boundary for YEAR is January 1, and even though the starting date is on a boundary, no boundaries are crossed in going from the first date to the second. Using the same logic, going from December 31, 2002, to January 1, 2003, *does* cross a year boundary and returns a 1. This is true even though, in the first case, there are 364 days between the dates and, in the latter case, only one day.

These intervals can be used "as is" or with multipliers such as two-year intervals, and they can be shifted so that the boundary is, for example, the seventh month of the year (July) instead of January 1.

When used with multi-intervals and shifted intervals, the INTCK function can become very complicated. A limited discussion of the finer points of the INTCK function follows the syntax and examples.

Syntax: INTCK('*interval<Multiple><.shift>*', *start-value*, *end-value*)

Intervals can be date units:

Interval	Description
DAY	Day
WEEK	Week
WEEKDAY	Each weekday (Monday to Friday, or any set of days you choose)
TENDAY	Ten-day period
SEMIMONTH	Two-week period
MONTH	Month
QTR	Quarter (Jan–Mar = 1, Apr–Jun = 2, etc.)
SEMIYEAR	Half year
YEAR	Year

Intervals can be time units:

Interval	Description
SECOND	Seconds
MINUTE	Minutes
HOUR	Hours

Intervals can be datetime units:

Interval	Description
DTDAY	Day
DTWEEK	Week
DTWEEKDAY	Each weekday (Monday to Friday)
DTTENDAY	Ten-day period
DTSEMIMONTH	Two-week period
DTMONTH	Month
DTQTR	Quarter (Jan–Mar = 1, Apr–Jun = 2, etc.)
DTSEMIYEAR	Half year
DTYEAR	Year

interval is one item from the preceding list, placed in quotation marks.

multiple is an optional modifier in the interval. You can specify multiples of an interval. For example, MONTH2 specifies two-month intervals; DAY50 specifies 50-day intervals.

.shift is an optional parameter that determines the starting point in an interval. For example, YEAR.4 specifies yearly intervals, starting from April 1. The shift value for single intervals is shown in the following table:

Shift value for SAS date and datetime values:

Interval	Shift Value
YEAR	Month
SEMIYEAR	Month
QTR	Month
MONTH	Month
SEMIMONTH	Semimonth*
TENDAY	Tenday
WEEKDAY	Day
WEEK	Day
DAY	Day

Shift value for SAS time intervals:

Interval	Shift Value
HOUR	Hour*
MINUTE	Minute*
SECOND	Second*

*Only multi-intervals of these intervals can be shifted.

For all multi-unit intervals except WEEK, SAS creates an interval starting from January 1, 1960. Multiple intervals are all shifted by the same unit as the non-multiple intervals (see lists above). So, YEAR4.24 specifies four-year intervals with the interval boundary at the beginning of the second year (January 1, 1962, January 1, 1966, etc.). MONTH4.2 indicates four-month intervals, with the boundary being the first day of the second month. See the following discussion on interval multipliers and shifted intervals.

Here are some examples of intervals:

Interval	Interpretation
YEAR	Each year
YEAR2	Every two years
YEAR.4	Each April
YEAR4.11	November, every four years
MONTH	Every month
MONTH4	Every four months
MONTH6.3	Every six months with boundaries at March and September
WEEK	Each week
WEEK2	Every two weeks
WEEK.4	Every week starting with Wednesday
WEEK2.4	Every two weeks starting with Wednesday
WEEKDAY	Five-day weeks with weekend days, Saturday and Sunday
WEEKDAY1W	Six-day weeks with weekend day, Sunday
WEEKDAY12W	Five-day weeks with weekend days, Sunday and Monday
HOUR	Every hour
HOUR4	Every four hours
HOUR8.7	Every eight hours with boundaries 6 AM, 2 PM, and 10 PM
DTMONTH	Every month (used with datetime values)

start-value is a SAS date, time, or datetime value.

end-value is a SAS date, time, or datetime value.

Examples

Function	Returns
INTCK('WEEK','16AUG2002'd,'24AUG2002'd)	1
INTCK('YEAR', '01JAN2002'd,'31DEC2002'd)	0
INTCK('YEAR', '01JAN2002'd,'02JAN2003'd)	1
INTCK('YEAR', '31DEC2002'd,'01JAN2003'd)	1
INTCK('QTR','01JAN2002'd,'01AUG2002'd)	2
INTCK('MONTH3','01JAN2002'd,'15APR2002'd)	1
INTCK('YEAR.7','05MAY2002'd,'15JUL2002'd)	1
INTCK('HOUR','06:01:00't,'07:23:15't)	1

See Program 4.9.

A Discussion of Interval Multipliers and Shifted Intervals

Some applications of interval multipliers are quite straightforward. For example, if you use YEAR2 as your interval, the intervals will be every two years. The value of

```
INTCK('YEAR2','15JAN2000'd,'21JAN2003'd)
```

is equal to 1 (one boundary, January 1, 2002, was crossed in going from January 15, 2000, to January 21, 2002). The reason that January 1, 2002, is a boundary is that the counting of boundaries goes back to January 1, 1960, which was an even number. Therefore, the boundaries will be even-numbered years.

You can shift some single intervals. For example, YEAR.7 indicates yearly intervals with the boundary being July 1 of every year. For the intervals of YEAR, SEMIYEAR, and QTR, the shift amount is months. For example, the value of

```
INTCK('YEAR.7','01JUN2000'd,'03JUL2002'd)
```

is equal to 3 (crossing boundaries at July 1, 2000, July 1, 2001, and July 1, 2002).

Shifting intervals that use multipliers is similar. For example, YEAR2.12 indicates two-year intervals, with boundaries at December of each two-year interval: December 1, 1960, December 1, 1962, etc. For example, the value of

```
INTCK('YEAR2.12','15JAN2000'd,'21JAN2003'd)
```

is equal to 2 (crossing the boundaries at December 1, 2000, and December 1, 2002).

Multi-month intervals are shifted by months, not weeks (since there is not an even number of weeks in a month). MONTH4.2 means four-month intervals with the boundary being the second month of each four-month period. By the way, the .2 does not mean "shift the boundary by 2 months." It means the boundary is the second month of each interval. As Charley Mullin says in his technical note: "The boundary is shifted TO an interval, not BY the interval."

The value of

```
INTCK('MONTH4.2','28JAN2003'd,'03JUL2003'd)
```

is equal to 2 (crossing the boundaries at February 1, 2003, and June 1, 2003).

WEEK and multi-week intervals present a special problem. For example, you might expect the value of

```
INTCK(WEEK,'01JAN1960'd,'04JAN1960'd)
```

to equal 0. However, it is equal to 1. The problem is that weekly intervals are counted every time a Sunday is crossed and January 1, 1960, is a Friday. The way that SAS decided to solve this problem was to start counting from Sunday in the same week of January 1, 1960, which is December 27, 1959. Going from January 1, 1960, to January 4, 1960, crosses a boundary (Sunday, January 3). This gets even more complicated when you are dealing with multi-week intervals.

As the default, the interval of WEEKDAY treats Saturday and Sunday as part of the preceding day. For example, the value of

```
INTCK('WEEKDAY','01JUN2003'd,'30JUN2003'd)
```

is equal to 21. June 1, 2003, is a Sunday, and June 30 falls on a Monday. The number of times you have crossed a boundary (a working day) is 21. (Please see a further discussion of the WEEKDAY interval below the June 2009 calendar later in this chapter.)

You can specify days other than Saturday and Sunday to be treated as weekend days. For example, if you had a six-day work week, with Sunday as the day off, you could indicate the interval as WEEKDAY1W. So, the value of

```
INTCK('WEEKDAY1W','01JUN2003'd,'30JUN2003'd)
```

is equal to 25 (Monday through Saturday for four weeks plus Monday, June 30).

If you were in the restaurant business and your restaurant was closed on Sunday and Monday, you would use the interval: WEEKDAY12W to compute the number of work days between two dates.

When you are computing the number of working days between two dates, it is important to know if the starting date is a working day or not. For example, take a look at the following calendar for June 2009:

JUNE						
S	M	T	W	T	F	S
	1	2	3	4	5	6
7	8	9	10	11	12	13
14	15	16	17	18	19	20
21	22	23	24	25	26	27
28	29	30				

The expression INTCK('weekday','08jun2009'd,'12jun2009'd) returns a 4, even though there are 5 working days in the week. Remember, you are counting how many boundaries are being crossed going from date 1 to date 2. Here, you cross a boundary on June 9, 10, 11, and 12 (4 boundaries).

Now, see what happens if you start on June 7 instead of June 8. The expression INTCK('weekday','07jun2009'd,'12jun2009'd) returns a 5, since you cross one boundary going from Sunday to Monday. By the way, if you computed the interval from June 6 to June 12, the result would also be a 5.

Notice that it doesn't matter if the ending date is a weekday or not. For example, INTCK('weekday','08jun2009'd,'13jun2009'd) or INTCK('weekday','08jun2009'd,'14jun2009'd) still returns a value of 4.

Function: INTNX

Purpose: To return the date after a specified number of intervals have passed.

Syntax: **INTNX('*interval*', *start-date*, *increment* <,'*alignment*'>)**

interval is one of the same values that are used with the INTCK function (placed in quotation marks).

start-date is a SAS date.

increment is the number of intervals between the start date and the date returned by the function.

alignment is an optional argument and has a value of BEGINNING (B), MIDDLE (M), END (E), or SAMEDAY(S). The default is BEGINNING. For example, if the interval is WEEK, an increment of 1 from January 1, 1960, with the default returns the date January 3, 1960 (a Sunday, the beginning of a boundary). The same date and interval with an alignment of MIDDLE returns the date January 6, 1960 (a Wednesday, the middle of the interval).

Examples

For these examples, DT1 = '01JAN1960:7:5:12'DT.
Note: Values in parentheses in the Returns column are the formatted values.

Function	Returns
INTNX('WEEK','01JAN1960'd,1)	2 (Sunday, Jan 3, 1960)
INTNX('WEEK','01JAN1960'd,1,'MIDDLE')	5 (Wednesday, Jan 6, 1960)
INTNX('WEEK.4','01JAN1960'd,1)	5 (Wednesday, Jan 6, 1960)
INTNX('WEEK2','01JAN1960'd,1)	9 (Sunday, Jan 10, 1960)
INTNX('QTR','01JAN2003'd,1)	15796 (Tuesday, April 1, 2003)
INTNX('YEAR.3','01JAN2003'd,1)	15765 (Saturday, March 1, 2003)
INTNX('YEAR.3','01JAN2003'd,2)	16131 (Monday, March 1, 2004)
INTNX('YEAR','01JUN2003'd,1)	16071 (Thursday, January 1, 2004)
INTNX('YEAR','01JUN2003'd,2)	16437 (Saturday, January 1, 2005)
INTNX('YEAR4.11','01JAN2003'd,1)	16376 (Monday, November 1, 2004)
INTNX('DTMONTH',DT1,3)	7862400 (01APR60:00:00:00)
INTNX('HOUR','9:15:09'T,2)	39600 (11:00:00)
INTNX('YEAR','15JAN1960'D,-1)	-365 (January 1, 1959)

Some examples demonstrating the SAMEDAY alignment

Date = '10May2005'd (Tuesday). Return values are formatted.

Function	Returns
INTNX('week',Date,1,'sameday')	17May2005 (Tuesday)
INTNX('month',Date,1,'sameday')	10Jun2005 (Friday)
INTNX('year',Date1,1,'sameday')	10May2006 (Wednesday)
INTNX('weekday','7May2005'd,1, 'sameday') Note: this is a Saturday	09May2005 (Monday)

Program 4.8: Demonstrating the INTNX function (with the SAMEDAY alignment)

```
***Primary functions: INTNX, WEEKDAY;
***Other functions: RANUNI, CEIL;

*A dentist wants to see each of his patients in six months for a followup
visit.  However, if the date in six months falls on a Saturday or Sunday,
he wants to pick a random day in the following week.;

Data dental;
   input Patno : $5. Visit_date : mmddyy10.;
   format Visit_date weekdate.;
datalines;
001 1/14/2009
002 1/17/2009
003 1/18/2009
004 1/19/2009
005 1/19/2009
006 1/20/2009
007 1/11/2009
008 1/17/2009
;
title "Listing of data set DENTAL";
proc print data=dental noobs;
run;
data followup;
   set dental;
   Six_months = intnx('month',Visit_date,6,'sameday');
   *Check if weekend;
   DayofWeek = weekday(six_months);
   *Keep track of actual day for testing purposes;
   Actual = Six_months;
   *If Sunday add random integer between 1 and 5;
   if DayofWeek = 1 then
      Six_months = Six_months + ceil(ranuni(0)*5);
   *If Saturday, add a random integer between 2 and 6;
   else if DayofWeek = 7 then
      Six_months = Six_months + ceil(ranuni(0)*5 + 1);
run;
title "Six Month Appointment Dates";
proc report data=followup nowd headline;
   columns Patno Visit_date Actual Six_months;
   define Patno / display "Patient Number" width=7;
   define Visit_date / display "Initial Date" width=15 format=weekdate.;
   define Actual / display "Actual Day" width=15 format=weekdate.;
   define Six_months / display "Six Month Appt." width=15
format=weekdate.;
run;
quit;
```

Explanation

The introduction of the SAMEDAY alignment greatly enhanced the usefulness of the INTNX function. If you used the INTNX function in the preceding program without the SAMEDAY alignment, all of the dentist's patients would be coming in on the first of each month! Not a great plan. By using the SAMEDAY alignment, the function returns a date six months ahead, but on the same day of the month. Since this date may be a Saturday or Sunday, adjustments need to be made. In this program, it was decided that if the six month date fell on a Saturday or Sunday, a random day in the following week was to be chosen.

The expression `ceil(ranuni(0)*5)` produces a random integer from 1 to 5; the expression `ceil(ranuni(0)*5 + 1)` produces a random integer from 2 to 6. For illustration purposes, the actual date six months from the visit date was not dropped from the data set so that you can see what happens if the follow-up date falls on a Saturday or Sunday. Here is the listing:

```
Six Month Appointment Dates

  Patient
  Number      Initial Date      Actual Day   Six Month Appt.

  001      Wed, Jan 14, 09  Tue, Jul 14, 09  Tue, Jul 14, 09
  002      Sat, Jan 17, 09  Fri, Jul 17, 09  Fri, Jul 17, 09
  003      Sun, Jan 18, 09  Sat, Jul 18, 09  Tue, Jul 21, 09
  004      Mon, Jan 19, 09  Sun, Jul 19, 09  Wed, Jul 22, 09
  005      Mon, Jan 19, 09  Sun, Jul 19, 09  Fri, Jul 24, 09
  006      Tue, Jan 20, 09  Mon, Jul 20, 09  Mon, Jul 20, 09
  007      Sun, Jan 11, 09  Sat, Jul 11, 09  Mon, Jul 13, 09
  008      Sat, Jan 17, 09  Fri, Jul 17, 09  Fri, Jul 17, 09
```

Using the INTNX Function to Determine Starting Boundaries for Multi-Day Intervals

Interval boundaries are straightforward for intervals such as years, quarters, and months. However, suppose you want to create 12-day intervals. How many 12-day intervals are there from January 1, 2004, to January 11, 2004? How many boundaries have you crossed? The problem here is that you have to realize that you start counting 12-day intervals from January 1, 1960, to determine where the boundaries are. Here's an easy way to see what date the counting starts on: use the INTNX function like this:

```
START_INTERVAL = INTNX('DAY12','01JAN2004'd,1)
```

The value is Saturday, January 10, 2004. So, in going from January 1, 2004, to January 11, 2004, you cross one boundary (January 10, 2004). To check, note that

```
INTCK('DAY12','01JAN2004'd,'11JAN2004'd)
```

is equal to 1.

Function: YRDIF

Purpose: To return the difference in years between two dates (includes fractional parts of a year).

Important note: About the time this book was being sent to the printer, it was discovered that the YRDIF function would sometimes return a value that was off by one day for certain date intervals. It appeared this error was related to leap years. However, this author believes that, even with this error, using the YRDIF function to compute ages (or any differences in years) is still more accurate than the older method of dividing the difference in years by 365.25. Future releases of SAS are expected to address this error with YRDIF. If you need to compute exact year differences and you are using a version of SAS that does not have the updated feature, you can use the INTCK function to accomplish your goals (see an illustration in this section).

Syntax: YRDIF(*start-date, end-date, 'basis'*)

start-date is a SAS date value.

end-date is a SAS date value.

basis is an argument that controls how SAS computes the result. The first value is used to specify the number of days in a month; the second value (after the slash) is used to specify the number of days in a year.

A value of 'ACT/ACT' (alias 'ACTUAL') uses the actual number of days in a month and the actual number of days in a year (either 365 or 366 days, depending on whether there are leap years involved). For certain industries, especially financial institutions, you can specify values for the number of days in the month and the number of days in the year. This is frequently done for interest calculations on bonds and other commodities. Other choices for *basis* are:

'30/360'	Uses 30-day months and 360-day years in the calculation.
'ACT/365'	Uses the actual number of days between the two dates, but uses 365-day years, even if a leap year is in the interval.
'ACT/360'	Uses the actual number of days between the two dates, but uses 360-day years.

Examples

Function	Returns
YRDIF('01JAN2002'd,'01JAN2003'd,'ACTUAL')	1
YRDIF('01JAN2002'd,01FEB2002'd,'ACT/ACT')	.0849
YRDIF('01FEB2002'd,01MAR2003'd,'ACTUAL')	1.9767
YRDIF('01JAN2002'd,'01JAN2003'd,'ACT/365')	1.0139

Program 4.9: Program to demonstrate the date interval functions

```
***Primary functions: INTCK, INTNX, YRDIF;

data period;
   set dates;
   Interval_month = intck('month',Date1,Date2);
   Interval_year  = intck('year',Date1,Date2);
   Year_diff      = yrdif(Date1,Date2,'actual');
   Interval_qtr   = intck('qtr',Date1,Date2);
   Next_month     = intnx('month',Date1,1);
   Next_year      = intnx('year',Date1,1);
   Next_qtr       = intnx('qtr',Date1,1);
   Six_month      = intnx('month',Date1,6);
   format Next: Six_month date9.;
run;
```

```
title "Listing of Data Set PERIOD";
proc print data=period heading=h;
   id date1 date2;
run;
```

Explanation

Before we discuss the date functions in this program, let me point out that the ID statement of PROC PRINT lists both DATE1 and DATE2 as ID variables. This allows the values to be repeated on the lower portion of the listing.

The interval functions can be somewhat confusing. It helps to keep in mind that the INTCK function counts how many times you cross a boundary going from the start date to the end date. The listing follows:

```
Listing of Data Set PERIOD


                      Interval_ Interval_  Year_  Interval_    Next_
    Date1     Date2    month      year     diff      qtr       month

01JAN1960 15JAN1960     0          0      0.03825     0     01FEB1960
02MAR1961 18FEB1962    11          1      0.96712     4     01APR1961
25DEC2000 03JAN2001     1          1      0.02461     1     01JAN2001
01FEB2002 31MAR2002     1          0      0.15890     0     01MAR2002



    Date1     Date2 Next_year      Next_qtr      Six_month

01JAN1960 15JAN1960 01JAN1961     01APR1960     01JUL1960
02MAR1961 18FEB1962 01JAN1962     01APR1961     01SEP1961
25DEC2000 03JAN2001 01JAN2001     01JAN2001     01JUN2001
01FEB2002 31MAR2002 01JAN2003     01APR2002     01AUG2002
```

Computing Exact Ages

If you are using a release of SAS that has not corrected the possible error in the YRDIF function, you can use the following SAS statement to compute AGE exactly (submitted by my friend Mike Zdeb):

```
Age_exact = floor((intck('month',DOB,Date)-(day(Date) < day(DOB))) / 12);
```

Function That Computes Dates of Standard Holidays

Function: **HOLIDAY**

Purpose: Returns a SAS date, given a holiday name and a year.

Syntax: HOLIDAY (*holiday*, *year*)

holiday is a holiday name (see list below).

year is a numeric variable or constant that represents the year.

Partial List of Holidays:

Christmas	Christmas day (December 25)
Columbus	Columbus day (2nd Monday in October)
Easter	Easter Sunday
Fathers	Father's Day (3rd Sunday in June)
Halloween	Halloween
Labor	Labor Day (1st Monday in September)
MLK	Martin Luther King Day (celebrated on Monday)
Memorial	Memorial Day (1st Monday in May)
Mothers	Mother's Day (2nd Sunday in May)
Newyear	New Year's Day (January 1)
Thanksgiving	Thanksgiving (4th Thursday in November)
USIndependence	July 4th holiday
USPresidents	President's Day (3rd Monday in February)
Veterans	Veterans Day (November 11)
VeteransUSG	Veterans Day (U.S. Government)

Examples

Function	Returns
HOLIDAY('Christmas',2009)	12/25/2009 (Friday)
HOLIDAY('USIndependence',2009)	7/4/2009 (Saturday)
HOLIDAY('VeteransUSG',2009)	11/11/2009 (Wednesday)
HOLIDAY('Easter',2009)	4/12/2009 (Sunday)
HOLIDAY('MLK',2009)	1/19/2009 (Monday)
HOLIDAY('Thanksgiving',2009)	11/26/2009 (Thursday)

Program 4.10: Demonstrating the HOLIDAY function

```
***Primary function: HOLIDAY;
***Other functions: WEEKDAY, INTCK;
data salary;
   H1 = holiday('Newyear',2005);
   if weekday(H1) = 7 then H1 = H1 + 2;
   else if weekday(H1) = 1 then H1 = H1 + 1;
   H2 = holiday('MLK',2005);
   H3 = holiday('USpresidents',2005);
   H4 = holiday('Easter',2005)-2;
   array H[4];
   First = '01Jan2005'd; *Saturday;
   Second = '31Mar2005'd; *Thursday;
   Work = intck('weekday',First,Second);
   /* if holiday falls between the First and Second date,
      decrement number of working days */
   do i = 1 to 4;
      if First le H[i] le Second then Work = Work - 1;
   end;
   Salary = 500 * Work;
   format First Second mmddyy10. Salary dollar10.;
   keep First Second Work Salary;
run;
title "Listing of SALARY";
proc print data=SALARY noobs;
run;
```

Explanation

In this program, you want to compute the number of working days between January 1, 2005 (Saturday) and March 31, 2005.

The INTCK function with the WEEKDAY interval computes the number of times you cross working day boundaries going from one date to another (with Monday through Friday being defined as the default working days).

Note: Since the starting date is a Saturday, you do not have to add one to the value returned, because going from a weekend day to Monday crosses a boundary. If the starting date were not on a weekend, you would need to add one to the variable WORK. (Please see the explanation following the June 2009 calendar earlier in this chapter.)

Next, you want to test if any of the holidays (New Year's Day, Martin Luther King's birthday, President's Day, or Easter) fall in that interval. In addition, since New Year's Day can fall on any day of the week, you use the WEEKDAY function to test if this holiday falls on a Saturday or Sunday. If so, your company gives its employees the following Monday off. Easter always falls on a Sunday, so employees are given the previous Friday (Good Friday) off.

An array is created to hold the four non-working days. Finally, you test if each of the four non-working days fall in the given interval. If so, you decrement the number of working days computed by the INTCK function.

Here is the listing:

```
Listing of SALARY

    First        Second     Work       Salary

01/01/2005   03/31/2005     60       $30,000
```

Functions That Work with Julian Dates

This group of functions involves Julian dates. Julian dates are commonly used in computer applications and represent a date as a two- or four-digit year followed by a three-digit day of the year (1 to 365 or 366, if it is a leap year). For example, January 3, 2003, in Julian notation would be either 2003003 or 03003. December 31, 2003 (a non-leap year) would be either 2003365 or 03365.

Function: DATEJUL

Purpose: To convert a Julian date into a SAS date.

Syntax: DATEJUL(*jul-date*)

jul-date is a numerical value representing the Julian date in the form *dddyy* or *dddyyyy*.

Examples

For these examples, JDATE = 1960123.

Function	Returns
DATEJUL(1960001)	0 (01JAN1960 formatted)
DATEJUL(2003365)	16070 (31DEC2003 formatted)
DATEJUL(JDATE)	122 (02MAY1960 formatted)

See Program 4.11.

Function: **JULDATE**

Purpose: To convert a SAS date into a Julian date.

Syntax: **JULDATE(*date*)**

date is a SAS date.

Examples

For these examples, DATE = '31DEC2003'D.

Function	Returns
JULDATE(DATE)	3365
JULDATE('01JAN1960'D)	60001
JULDATE(122)	60123

See Program 4.11.

Function: JULDATE7

Purpose: To convert a SAS date into seven-digit Julian date.

Syntax: JULDATE7 (*date*)

date is a SAS date.

Examples

For these examples, DATE = '31DEC2003'D.

Function	Returns
JULDATE7(DATE)	2003365
JULDATE7('01JAN1960'D)	1960001
JULDATE7(122)	1960123

Program 4.11: Demonstrating the three Julian date functions

```
    ***Primary functions: DATEJUL, JULDATE, and JULDATE7.;

***Note: option YEARCUTOFF set to 1920;
options yearcutoff = 1920;
data julian;
   input Date : date9. Jdate;
   Jdate_to_sas = datejul(Jdate);
   Sas_to_Jdate = juldate(Date);
   Sas_to_jdate7 = juldate7(Date);
   format Date Jdate_to_sas mmddyy10.;
datalines;
01JAN1960 2003365
15MAY1901 1905001
21OCT1946 5001
;

title "Listing of Data Set JULIAN";
proc print data=julian noobs;
   var Date Sas_to_jdate Sas_to_jdate7 Jdate Jdate_to_sas;
run;
```

Explanation

It is important to realize that Julian dates without four-digit years will be converted to SAS dates, based on the value of the YEARCUTOFF system option. To avoid any problems, it is best to use seven-digit Julian dates. The listing is shown next:

```
Listing of Data Set JULIAN

             Sas_to_    Sas_to_                    Jdate_
    Date      Jdate      jdate7      Jdate         to_sas

01/01/1960    60001     1960001     2003365      12/31/2003
05/15/1901   1901135    1901135     1905001      01/01/1905
10/21/1946    46294     1946294        5001      01/01/2005
```

Chapter 5

Array Functions

Introduction

Although there are only three array functions, chances are you will need to use some or all of them if you use arrays in your programs. They are all used either to determine the number of elements in, or the upper and lower bounds of, a SAS array.

Function:　DIM

When you define an array, you can either enter the number of elements in the array in parentheses (or square or curly brackets) following the array name or you can use an asterisk (*) to indicate that you don't care to count the number of elements (perhaps you are "counting challenged"). SAS programmers often use an asterisk in place of the number of array elements when the list of variables is very long or where they are using keywords such as _CHARACTER_ or _NUMERIC_ in place of the list of variables. However, when you want to use a DO loop to process all the elements of the array, you need to know the number of elements. Here is where the DIM function comes in handy. It returns the number of elements in an array, given the array name as its argument.

Purpose: To determine the number of elements in an array.

Syntax: DIM(*array-name*<,*dimension*>)

or

DIM*n*(*array-name*) where *n* is a dimension of a multidimensional array

array-name is the name of the SAS array for which you want to determine the number of elements.

Examples

For these examples, four arrays are defined as follows:

```
ARRAY ONE[*] X1-X3;
ARRAY TWO[3:6] A1 A2 A3 A4;
ARRAY THREE[2,3] X1-X6;
ARRAY MULT(5,3) M1-M15;
```

Function	Returns
DIM(ONE)	3
DIM(TWO)	4
DIM(THREE)	2 (number of elements in the 1st dimension)
DIM(MULT)	5 (number of elements in the 1st dimension)
DIM(MULT,2)	3 (number of elements in the 2nd dimension)
DIM2(MULT)	3 (number of elements in the 2nd dimension)

Program 5.1: Setting all numeric values of 999 to missing and all character values of 'NA' to missing

```
***Primary function: DIM
***Other function: UPCASE;

data mixed;
   input X1-X5 A $ B $ Y Z;
datalines;
1 2 3 4 5 A b 6 7
2 999 6 5 3 NA C 999 10
5 4 3 2 1 na B 999 999
;
data array_1;
   set mixed;
```

```
     array nums[*] _numeric_;
     array chars[*] _character_;
     do i = 1 to dim(nums);
         if nums[i] = 999 then nums[i] = .;
     end;

   do i = 1 to dim(chars);
      if upcase(chars[i]) = 'NA' then chars[i] = ' ';
      end;
      drop i;
   run;
   title "Listing of Data Set ARRAY_1";
   proc print data=array_1 noobs;
   run;
```

Explanation

The keys to this program are the two keywords _NUMERIC_ and _CHARACTER_. These refer to all numeric and character variables defined at that point in the DATA step. Since the two ARRAY statements follow a SET statement, the _NUMERIC_ keyword will represent all the numeric variables in the MIXED data set. In a similar manner, _CHARACTER_ will represent all the character variables in the MIXED data set. (Remember that an array must consist of all character or all numeric variables.) Note that if these two ARRAY statements were placed before the SET statement, the two arrays would have no elements. Since the asterisk (*) is used to represent the number of elements in the two arrays (because it could be inconvenient to obtain these numbers or because these numbers might change), it is convenient to use the DIM function to determine this number.

The first DO loop starts from the number one and ends with the number of numeric variables in the MIXED data set. Inside the loop, you check to see if the value is equal to 999, in which case you then set the value to a SAS missing value. In a similar manner, the next DO loop starts from the number one and ends with the number of character variables in the data set MIXED. Notice the use of the UPCASE function to convert any lowercase values to uppercase. This results in any uppercase or lowercase values of 'NA' being set to missing values. Here is a listing of data set ARRAY_1:

```
Listing of Data Set ARRAY_1

X1    X2    X3    X4    X5    A    B    Y    Z

1     2     3     4     5     A    b    6    7
2     .     6     5     3          C    .    10
5     4     3     2     1          B    .    .
```

Program 5.2: Creating a macro to compute and print out the number of numeric and character variables in a SAS data set

```
***Primary function: DIM;

%macro count(Dsn= /*Data set name */);
   title1 "*** statistics for data set &dsn ***";
   data _null_;
      if 0 then set &Dsn;
      array nums[*] _numeric_;
      array chars[*] _character_;
      n_nums = dim(nums);
      n_chars = dim(chars);
      file print;
      put / "There are " n_nums "numeric variables and "
         n_chars "character variables";
   stop;
   run;
%mend count;
```

Explanation

This macro uses an interesting trick to determine the number of numeric and character variables in a SAS data set—that is, the strange-looking conditional SET statement. The SET statement follows an IF statement (IF 0) that is never true, so the data values are never read. You would think that nothing useful could come from such a statement. However, during the compile stage, SAS performs certain operations, one being the evaluation of the list of numeric and character variables in the data set listed in the SET statement. (Veteran SAS programmers often use the trick of a conditional SET statement to determine the number of observations in a SAS data set by using the NOBS= option in the SET statement.) So, even though the SET statement is never executed in the execution stage, the two arrays defined by the _NUMERIC_ and _CHARACTER_ keywords are properly assigned, and the DIM function determines the number of elements in each of the two arrays. To test this macro, we can call it like this:

```
%COUNT(Dsn=array_1)
```

with the result listed next:

```
*** statistics for data set array_1 ***

There are 7 numeric variables and 2 character variables
```

Using the DIM Function with Multidimensional Arrays

When you use the DIM function with a multidimensional array, you have two choices:

First, you can append the number of the dimension that you are checking to the function name. For example, you would use DIM2 (*array-name*) to determine the number of elements in the second dimension of an array.

Second, you can enter the array dimension as the second argument of the DIM function. For example, you would use DIM (Mult, 2) to determine the number of elements in the second dimension of an array.

Functions: HBOUND and LBOUND

Purpose: To determine the lower and upper bounds of the array when the array bounds do not run from 1 to *n*.

Syntax: HBOUND (*array-name*)   or   LBOUND (*array-name*)

array-name is the name of any previously defined SAS array.

When you index your array starting from the number one (the default), the DIM function works just fine. However, if you explicitly set your subscripts to start from a number other than one (using a colon between the lower and upper bounds of the array in the parentheses following the array name), the HBOUND and LBOUND functions become useful. Remember that the DIM function returns the number of elements, not the upper bound.

Program 5.3: Determining the lower and upper bounds of an array where the array bounds do not start from one

```
***Primary functions: LBOUND and HBOUND;

data array_2;
   array Income[1990:1995] Income1990-Income1995;
   array Tax[1990:1995] Tax1990-Tax1995;
   input Income1990 - Income1995;
   do Year = lbound(Income) to hbound(Income);
      Tax[year] = .25 * Income[year];
   end;
   format Income: Tax: dollar8.;
   drop Year;
```

```
datalines;
50000 55000 57000 66000 65000 68000
;
title "Listing of Data Set ARRAY_2";
proc print data=array_2 noobs;
run;
```

Explanation

This programmer found it convenient to use YEAR as the subscript in the two arrays in the program. Notice the colon separating the lower and upper bounds of these two arrays. In this case, the two bounds are obvious, but we will use the two bound functions anyway to demonstrate them. Here, the expression LBOUND(INCOME) will return the value 1990 and the expression HBOUND(INCOME) will return the value 1995. The FORMAT statement uses a little known, but useful SAS feature. The colon following the two variable names in this statement acts as a "wildcard" character. For example, INCOME: refers to all variables in the data set that begin with the letters "INCOME". Here is a listing of data set ARRAY_2:

```
Listing of Data Set ARRAY_2

Income1990   Income1991   Income1992   Income1993   Income1994   Income1995

 $50,000      $55,000      $57,000      $66,000      $65,000      $68,000


Tax1990      Tax1991      Tax1992      Tax1993      Tax1994      Tax1995

$12,500      $13,750      $14,250      $16,500      $16,250      $17,000
```

Before we leave the topic of array bounds, it is instructive to point out that when you supply your own array bounds, you would most likely want to be sure that you also supply a variable list. If you do not supply this list, SAS will assign variable names starting from one. For example, if you declare an array like this:

```
array Income[1990:1995];
```

SAS will reference variables INCOME1 to INCOME6.

C h a p t e r   6

Truncation Functions

Introduction

This chapter covers five functions that involve rounding or truncation. TRUNC allows you to perform logical comparisons on numerical values that were stored with fewer than 8 bytes. There is also a novel use of the ROUND function to place values into groups, such as an age group.

There is a group of functions (CEILZ, FLOORZ, INTZ, and ROUNDZ) similar to the ones in this chapter, except that they do not "fuzz" the result. Fuzzing the result means to return an integer value if the result is within 10^{-12} of an integer.

Functions That Round and Truncate Numerical Values

Function: **CEIL**

Purpose: To round up to the next largest integer.

Syntax: CEIL(*numeric-value*)

numeric-value is any SAS numeric variable, expression, or numerical constant. For negative arguments, the returned value is an integer between the number and zero (see the following examples). For positive arguments, the function returns just the integer portion of the number (the same as the INT function). Values within 10^{-12} of the value of an integer will return the integer.

Examples

Function	Returns
CEIL(1.2)	2
CEIL(1.8)	2
CEIL(-1.2)	-1
CEIL(-1.8)	-1

Program 6.1: Using the CEIL function to round up a value to the next penny

```
***Primary function CEIL;

data roundup;
   input Money @@;
   Up = ceil(100*Money)/100;
datalines;
123.452 45 1.12345 4.569
;
title "Listing of Data Set ROUNDUP";
proc print data=roundup noobs;
run;
```

Explanation

You can use the CEIL function in cases where you want the next largest integer. Remember that for negative arguments, this function is equivalent to the INT function: the values will be rounded up (i.e., closer to zero). Here is the listing of data set ROUNDUP:

```
Listing of Data Set ROUNDUP

 Money      Up

123.452    123.46
 45.000     45.00
  1.123      1.13
  4.569      4.57
```

Function: **FLOOR**

Purpose: To round down to the next smallest integer. This is sometimes used for ages, where you want a person's age as of his or her last birthday.

Syntax: FLOOR(*numeric-value*)

numeric-value is any SAS numeric variable, expression, or numerical constant. For positive arguments, this function is equivalent to the INT function. Values within 10^{-12} of an integer will return the value of the integer. For negative arguments, this function returns the next lowest negative number (see the following examples):

Examples

Function	Returns
FLOOR(1.2)	1
FLOOR(1.8)	1
FLOOR(-1.2)	-2
FLOOR(-1.8)	-2

Program 6.2: Computing a person's age as of his or her last birthday (rounding down)

```
***Primary function: FLOOR;
***Other function: YRDIF;

data fleur;
   input @1 DOB mmddyy10.;
   Age = yrdif(DOB,'01jan2003'd,"actual");
   Age_floor = floor(Age);
   ***note: since age is positive, the int function is equivalent;
   format DOB mmddyy10.;
datalines;
10/21/1946
05/25/2000
;
title "Listing of Data Set FLEUR";
proc print data=fleur noobs;
run;
```

Explanation

AGE is computed by using the YRDIF function (see Chapter 4, "Date and Time Functions"). Since the result may include a fractional part of a year, the FLOOR function is used to return the value of the next smallest integer, the person's age as of his or her last birthday. Note that since AGE will always be a positive number, the INT function would give equivalent results and could be used here. Here is a listing of data set FLEUR:

```
Listing of Data Set FLEUR

                       Age_
        DOB      Age   floor

10/21/1946   56.1973     56
05/25/2000    2.6038      2
```

Function: INT

Purpose: To remove the fractional part of a number.

Syntax: INT (*numeric-value*)

numeric-value is any SAS numeric variable, expression, or numerical constant. For positive arguments, this function is equivalent to the FLOOR function. Values within 10^{-12} of the value of an integer will return the value of the integer. For negative arguments, this function returns the next highest negative number (see examples below):

Examples

Function	Returns
INT(1.2)	1
INT(1.8)	1
INT(-1.2)	-1
INT(-1.8)	-1

Program 6.3: Using the INT function to compute age as of a person's last birthday

```
***Primary function: INT;

data ages;
   informat DOB mmddyy10.;
   input DOB @@;
   Age = ('01jan2003'd - DOB)/365.25;
   Age_int = int(age);
   format DOB mmddyy10.;
datalines;
10/21/1946 11/12/1956 6/7/2002 12/20/1966 3/6/1930 5/8/1980
;
title "Listing of Data Set AGES";
proc print data=ages noobs;
run;
```

Explanation

This program is similar to Program 6.2. In this example, AGE is computed directly, without the use of the YRDIF function, and the INT function is used instead of the FLOOR function to drop the fractional part of the age.

```
Listing of Data Set AGES

      DOB      Age     Age_int

10/21/1946   56.1973      56
11/12/1956   46.1366      46
06/07/2002    0.5699       0
12/20/1966   36.0329      36
03/06/1930   72.8247      72
05/08/1980   22.6503      22
```

Function: **ROUND**

Purpose: To round a numerical value. The rounding can be to the nearest integer or to any desired value, such as to the nearest tenth or the nearest 10.

Syntax: ROUND(*numeric-value* <, *round-off-unit*>)

numeric-value is any SAS variable, numeric expression, or numeric constant.

round-off-unit is a value to determine how the rounding should be done. If *round-off-unit* is omitted, the default round-off unit is 1. A value of .1 will round to the nearest tenth. A value of 10 will round to the nearest 10.

Examples

Function	Returns
ROUND(1.2)	1
ROUND(1.8)	2
ROUND(3.14159,.01)	3.14
ROUND(23.1,5)	25
ROUND(-1.2)	-1

Program 6.4: Rounding students' grades several ways

```
***Primary function: ROUND;
***Other function: MEAN;

data scores;
   input ID Test1-Test3;
   Test_ave = mean(of Test1-Test3);
   Round_one = round(Test_ave);
   Round_tenth = round(Test_ave,.1);
   Round_two = round(Test_ave,2);
datalines;
1  100 95 95
2  78 79 88
;
title "Listing of Data Set SCORES";
proc print data=scores noobs;
   id ID;
   var Test_ave Round:;
run;
```

Explanation

This program demonstrates several variations on the ROUND function. The first line using the ROUND function does not include the optional second argument, so the value is rounded to the nearest integer. The next line rounds the test average to the nearest tenth. Finally, the third use of the ROUND function rounds the test average to the nearest two points. The following listing shows the results of the different round-off units:

```
Listing of Data Set SCORES

             Round_    Round_    Round_
ID   Test_ave   one     tenth     two

 1    96.6667    97      96.7      96
 2    81.6667    82      81.7      82
```

Program 6.5: Using the ROUND function to group ages into 10-year intervals

```
***Primary function: ROUND;
***Other function: YRDIF;

data decades;
   informat DOB mmddyy10.;
   input DOB @@;
   Age = yrdif(DOB,'01jan2003'd,"actual");
   Decade = round(Age + 5., 10);
   format DOB mmddyy10.;
datalines;
10/21/1946 11/12/1956 6/7/2002 12/20/1966 3/6/1930 5/8/1980
11/11/1998 10/21/1990 5/5/1994 10/21/1992
;
title "Listing of Data Set DECADES";
proc print data=decades noobs;
run;
```

Explanation

This is a somewhat novel use of the ROUND function. You have to take great care when you use this function to group values. In this program, the round-off unit is 10. If you didn't add 5 to the AGE, ages from 0 to 5 would be set to 0, greater than 5 to less than 15 would be rounded to 10, and so forth. If, instead, you want the 0–10 group to be set to 10, 10–20 to be set to 20, etc., you can add 5 to the value before you do the rounding. A more direct method, using a user-defined format and a PUT function (or a series of IF and ELSE IF statements) might be less prone to error than the method presented here. But, then again, this method requires only a single line of code. The following listing shows how the rounding was performed:

```
Listing of Data Set DECADES

       DOB       Age       Decade

10/21/1946    56.1973       60
11/12/1956    46.1366       50
06/07/2002     0.5699       10
12/20/1966    36.0329       40
03/06/1930    72.8247       80
05/08/1980    22.6503       30
11/11/1998     4.1397       10
10/21/1990    12.1973       20
05/05/1994     8.6603       10
10/21/1992    10.1967       20
```

Demonstrating the Difference between Various Truncation Functions

The following program (thanks to Mike Zdeb) demonstrates how the various truncation functions treat positive and negative values:

Program 6.6: Demonstrating the various truncation functions

```
***Primary functions: CEIL, FLOOR, INT, and ROUND;

data truncate;
   input x @@;
   Ceil = ceil(x);
   Floor = floor(x);
   Int = int(x);
   Round = round(x);
datalines;
7.2 7.8 -7.2 -7.8
;
title "Listing of Data Set TRUNCATE";
proc print data=truncate noobs;
run;
```

Explanation

The following listing shows how the various truncation functions work with positive and negative numbers:

```
Listing of Data Set TRUNCATE

  x     Ceil    Floor    Int    Round

 7.2     8       7        7       7
 7.8     8       7        7       8
-7.2    -7      -8       -7      -7
-7.8    -7      -8       -7      -8
```

Function That Returns SAS Numerical Values Stored in Fewer than 8 Bytes

Function: TRUNC

Purpose: To allow you to make comparisons between numerical constants and SAS numerical values stored in fewer than 8 bytes. Note that this is necessary only for non-integer values. Be sure to refer to the system companion book for a chart that tells you the largest integer that can be expressed in *n* bytes of storage.

Syntax: TRUNC(*numeric-value, length*)

numeric-value is any SAS variable, numeric expression, or numeric constant.

length is an integer between 3 and 8, representing the number of bytes.

Examples

Function	Returns
TRUNC(3.,3)	3.0000000000000
TRUNC(3.,4)	3.0000000000000
TRUNC(3.,8)	3.0000000000000
TRUNC(3.1,3)	3.0996093750000
TRUNC(3.1,4)	3.0999984741211
TRUNC(3.1,8)	3.1000000000000

Program 6.7: Logical comparisons with numbers stored with fewer than 8 bytes of precision

```
***Primary function: TRUNC;

data test;
   length x4 4 x8 8;
   input x4 x8;
datalines;
1.234 1.234
;
data trunctest;
   set test;
   if x8 = 1.234 then Compare_x8 = 'True ';
   else Compare_x8 = 'False';
   if x4 = 1.234 then Compare_x4 = 'True ';
   else Compare_x4 = 'False';
   if x4 = trunc(1.234,4) then Compare_trunc_x4 = 'True ';
   else Compare_trunc_x4 = 'False';
run;

title "Listing of Data Set TRUNTEST";
proc print data=trunctest noobs;
run;
```

Explanation

To understand what is going on here, you need to remember that all numerical values, regardless of their storage length, are expanded to 8 bytes when they are brought into the PDV. When this happens, you lose precision. For example, if you store a numeric value in 4 bytes and it is brought into the PDV, SAS adds 4 bytes of zeros. If you then try to compare this value to a numerical constant, the values may not agree. In this program, the variables X4 and X8 are read from data set TEST. The value of X8 agrees with the numerical constant and COMPARE_X8 is 'TRUE'. However, when you compare the stored value of X4 against the numerical constant, they do not agree. So, when you check the value of numerical variables stored in fewer than 8 bytes, you should use the TRUNC function to create a numerical constant with the same number of bytes as the stored value.

Note that if your values are integers, you do not have to use the TRUNC function to make your comparisons.

See the listing below to confirm this:

```
Listing of Data Set TRUNTEST

                    Compare_    Compare_    Compare_
    x4        x8       x8          x4        trunc_x4

1.23400    1.234      True       False        True
```

Chapter 7

Descriptive Statistics Functions

Introduction

The name of this chapter, "Descriptive Statistics Functions," is somewhat misleading. Although some of the functions, such as MEAN (average) and STD (standard deviation), can be thought of as statistical, many of these functions have extremely useful, everyday, non-statistical uses. So, don't be scared off. As is true throughout the book, I've used the same function categories as the SAS documentation to avoid confusion. You will find some of these functions indispensable in your SAS programming.

Functions That Determine the Number of Missing and Non-missing Values in a List of SAS Variables

Function: **N**

Purpose: To determine the number of non-missing values in a list of numeric values.

Syntax: N(<of> *numeric-values*)

numeric-values is either a list of numeric values (numeric variables or numbers), separated by commas, or a list of variables in the form BASE*n*–BASE*m* (e.g., X1–X5). If the latter style is used, you must place the word "of" before the list. The reason for this is that without the word "of," SAS interprets a list such as X1–X5 as the value of X1 minus the value of X5.

Examples

For these examples, X = 5, Y = ., Z = 7, X1 = 1, X2 = 2, X3 = 3, X4 = .

Function	Returns
N(X,Y,Z)	2
N(OF X1-X4)	3
N(10,20,. ,30,40,50)	5
N(OF X1-X4,Y,Z)	4

Program 7.1: Using the N function to determine the number of non-missing values in a list of variables

---

```
***Primary functions: N, MEAN;

data quiz;
   input Quiz1-Quiz10;
   ***compute quiz average if 8 or more quizzes taken;
   if n(of Quiz1-Quiz10) ge 8 then Quiz_ave = mean(of Quiz1-Quiz10);
datalines;
90 88 79 100 97 96 94 95 93 88
60 90 66 77 . . . 88 84 86
90 . . 90 90 90 90 90 90 90
;
title "Listing of Data Set QUIZ";
proc print data=quiz noobs heading=h;
run;
```

Explanation

This program uses the N function to determine the number of non-missing quiz scores. If the number is eight or more, an average is computed. Otherwise, QUIZ_AVE will be a missing value. In the following listing, notice the missing value for QUIZ_AVE in the second observation.

```
Listing of Data Set QUIZ

Quiz1 Quiz2 Quiz3 Quiz4 Quiz5 Quiz6 Quiz7 Quiz8 Quiz9 Quiz10 Quiz_ave

  90    88    79   100    97    96    94    95    93    88      92
  60    90    66    77     .     .     .    88    84    86       .
  90     .     .    90    90    90    90    90    90    90      90
```

Function: NMISS

Purpose: To determine the number of missing values in a list of numeric values.

Syntax: NMISS(<of> *numeric-values*)

numeric-values is either a list of numeric values (numeric variables or numbers), separated by commas or a list of variables in the form BASE*n*–BASE*m* (e.g., X1–X5). If the latter style is used, you must place the word "of" before the list.

Examples

For these examples, X = 5, Y = ., Z = 7, X1 = 1, X2 = 2, X3 = 3, X4 = .

Function	Returns
NMISS(X,Y,Z)	1
NMISS(OF X1-X4)	1
NMISS(10,20,. ,30,40,50)	1
NMISS(OF X1-X4,Y,Z)	2

Program 7.2: Computing a SUM and MEAN of a list of variables, only if there are no missing values

```
***Primary functions: NMISS, MEAN and SUM
***Other function: N;

data nomiss;
   input X1-X3 Y Z;
   if nmiss(of X1-X3,Y,Z) eq 0 then do;

   ***An alternative statement is:
   if n(of X1-X3,Y,Z) eq 5 then do;

      Sum_vars = sum(of X1-X3,Y,Z);
      Average = mean(of X1-X3,Y,Z);
   end;
datalines;
1 2 3 4 5
9 . 8 7 6
8 8 8 8 8
;
title "Listing of Data Set NOMISS";
proc print data=nomiss noobs;
run;
```

Explanation

In this example, you want to compute a SUM and MEAN only if there are no missing values in the list of variables. If the NMISS function returns a 0, you know that there are no missing values in your list of numeric values, and you can compute the sum and mean of these numeric values. If there are any missing values, NMISS will be greater than 0 and the value of SUM_VARS and AVERAGE will remain missing. You could, alternatively, use a regular assignment statement to compute the SUM and MEAN, but if there is a long list of variables, this could get tedious. A listing of data set NOMISS is shown next:

```
                        Listing of Data Set NOMISS

X1    X2    X3    Y    Z    Sum_vars    Average

 1     2     3    4    5       15          3
 9     .     8    7    6        .          .
 8     8     8    8    8       40          8
```

Functions That Compute Sums, Means, and Medians

Function: MEAN

Purpose: To compute the mean (average) of the non-missing values of a list of numeric values.

Syntax: MEAN(<of> *numeric-values*)

numeric-values is either a list of numeric values (numeric variables or numbers), separated by commas, or a list of variables in the form BASE*n*–BASE*m* (e.g., X1–X5). If the latter style is used, you must place the word "of" before the list. Note that this function computes the mean only of the non-missing values in the list. If all numeric values in the list have missing values, the result is a missing value.

Examples

For these examples, $X = 5$, $Y = .$, $Z = 7$, $X1 = 1$, $X2 = 2$, $X3 = 3$, $X4 = .$

Function	Returns
MEAN(X,Y,Z)	6
MEAN(OF X1-X4)	2
MEAN(10,20,. ,30,40,50)	30
MEAN(OF X1-X4,Y,Z)	3.25

See Program 7.3 for an example.

Function: **MEDIAN**

Purpose: To compute the median of the non-missing values of a list of numeric values.

Syntax: `MEDIAN(<of> numeric-values)`

numeric-values is either a list of numeric values (numeric variables or numbers), separated by commas, or a list of variables in the form BASE*n*–BASE*m* (e.g., X1–X5). If the latter style is used, you must place the word "of" before the list. Note that this function computes the median only of the non-missing values in the list. If all numeric values in the list have missing values, the result is a missing value.

Examples

For these examples, $X = 5$, $Y = .$, $Z = 7$, $X1 = 1$, $X2 = 2$, $X3 = 3$, $X4 = .$

Function	Returns
`MEDIAN(X,Y,Z)`	6
`MEDIAN(OF X1-X4)`	2
`MEDIAN(10,20,. ,30,40,50)`	30
`MEDIAN(OF X1-X4,Y,Z)`	2.5

See Program 7.3 for an example.

Function: **SUM**

Purpose: To compute the sum of the non-missing values in a list of numeric values.

Syntax: `SUM(<of> numeric-values)`

numeric-values is either a list of numeric values (numeric variables or numbers), separated by commas, or a list of variables in the form BASE*n*–BASE*m* (e.g., X1–X5). If the latter style is used, you must place the word "of" before the list. Note that this function computes the sum only of the non-missing values in the list. If all numeric values in the list have missing values, the result is a missing value. A useful trick is to add a 0 in the list of numeric values if you want the function to return a 0 if all the values are missing. For example: `TOTAL = SUM(0, OF COST1-COST5).`

Examples

For these examples, $X = 5$, $Y = .$, $Z = 7$, $X1 = 1$, $X2 = 2$,
$X3 = 3$, $X4 = .$

Function	Returns
SUM(X,Y,Z)	12
SUM(OF X1-X4)	6
SUM(10,20,. ,30,40,50)	150
SUM(0,Y,X4)	0

Program 7.3: Computing a mean, median, and sum of eight variables, only if there are at least six non-missing values

```
***Primary functions: N, MEAN, MEDIAN, SUM;

data score;
   input @1 (Item1-Item8)(1.);
   if n(of Item1-Item8) ge 6 then do;
      Mean = mean(of Item1-Item8);
      Median = median(of Item1-Item8);
      Sum = sum(of Item1-Item8);
   end;
datalines;
12345678
1.3.5678
1...5678
;
title "Listing of SCORE";
proc print data=score noobs;
run;
```

Explanation

The combination of the N function with several of the descriptive statistics functions is particularly useful. Since the MEAN, MEDIAN, and SUM functions ignore missing values, you may want to use the N function to determine if there is a minimum number of values before you do your calculations. In the preceding program, the mean, median, and sum are computed only if the number of non-missing values is six or more. A listing of the resulting data set follows. Notice that the mean, median, and sum have missing values in the third observation.

```
Listing of SCORE

Item1 Item2 Item3 Item4 Item5 Item6 Item7 Item8 Mean Median Sum

  1     2     3     4     5     6     7     8    4.5   4.5    36
  1     .     3     .     5     6     7     8    5.0   5.5    30
  1     .     .     .     5     6     7     8     .     .      .
```

Functions That Compute the Spread or Dispersion of Data Values

Function: **RANGE**

Purpose: To compute the range (distance between the highest and lowest value) in a list of numeric values.

Syntax: **RANGE** (**<of>** *numeric-values*)

numeric-values is either a list of numeric values (numeric variables or numbers), separated by commas, or a list of variables in the form BASE*n*–BASE*m* (e.g., X1–X5). If the latter style is used, you must place the word "of" before the list. The reason for this is that without the word "of," SAS interprets a list such as X1–X5 as the value of X1 minus the value of X5.

Examples

For these examples, $X = 5$, $Y = .$, $Z = 7$, $X1 = 1$, $X2 = 2$, $X3 = 3$, $X4 = .$

Function	Returns
RANGE(X,Y,Z)	2
RANGE(OF X1-X4)	2
RANGE(10,20,. ,30,40,50)	40
RANGE(OF X1-X4,Y,Z)	6

See Program 7.4 for an example.

Function: IQR

Purpose: To compute the interquartile range (distance between the 25th percentile and the 75th percentile) in a list of numeric values.

Syntax: `IQR(<of> numeric-values)`

`numeric-values` is either a list of numeric values (numeric variables or numbers), separated by commas, or a list of variables in the form BASE*n*– BASE*m* (e.g., X1–X5). If the latter style is used, you must place the word "of" before the list. The reason for this is that without the word "of," SAS interprets a list such as X1–X5 as the value of X1 minus the value of X5.

Examples

For these examples, $X = 5$, $Y = .$, $Z = 7$, $X1 = 1$, $X2 = 2$, $X3 = 3$, $X4 = .$

Function	Returns
`IQR(X,Y,Z)`	2
`IQR(OF X1-X4)`	2
`IQR(10,20,. ,30,40,50)`	20
`IQR(OF X1-X4,Y,Z)`	3.5

See Program 7.4 for an example.

Function: STD

Purpose: To compute the standard deviation of a series of variables in a single observation.

Syntax: `STD(<of> numeric-values)`

`numeric-values` is either a list of numeric values (numeric variables or numbers), separated by commas, or a list of variables in the form BASE*n*– BASE*m* (e.g., X1–X5). If the latter style is used, you must place the word "of" before the list. Note that this function computes the standard deviation using only the non-missing values in the list. If there are fewer than two non-missing values in the list, the result is a missing value.

Examples

For these examples, $X = 5$, $Y = .$, $Z = 7$, $X1 = 1$, $X2 = 2$, $X3 = 3$, $X4 = .$

Function	Returns
STD(X,Y,Z)	1.414
STD(OF X1-X4)	1
STD(10,20,. ,30,40,50)	15.811
STD(OF X1-X4,Y,Z)	2.630

Program 7.4: Computing a range, interquartile range, and standard deviation for each subject

```
***Primary functions: RANGE, IQR, and STD;

data home_on_the_range;
   input Subject X1-X10;
   Range = range(of X1-X10);
   IQR = iqr(of X1-X10);
   SD = std(of X1-X10);
datalines;
1  1 2 3 4 5 6 7 8 9 10
2  9 7 4 1 15 0 . 2 7 4
3  1 3 5 7 9 11 13 15 20 100
;
title "Listing of Data Set HOME_ON_THE_RANGE";
proc print data=home_on_the_range;
   id Subject;
run;
```

Explanation

This program computes the range (RANGE), interquartile range (IQR), and sample standard deviation (STD) for each subject. Notice that the differences among the three measures of range vary widely, especially when there is an outlier in the data, as with subject three (value of 100). In this case, the inclusion of the outlier greatly inflates the range and standard deviation but has no effect on the IQR. Note: if you replaced the value of 100 with a value of 1000, the IQR would not change.

```
Listing of Data Set HOME_ON_THE_RANGE

Subject  X1  X2  X3  X4  X5  X6  X7  X8  X9  X10  Range  IQR    SD

   1      1   2   3   4   5   6   7   8   9   10     9     5   3.0277
   2      9   7   4   1  15   0   .   2   7    4    15     5   4.6667
   3      1   3   5   7   9  11  13  15  20  100    99    10  29.2392
```

Functions That Determine the Ordering of Data Values

Function: MIN

Purpose: To determine the smallest non-missing value in a list of numeric values.

Syntax: `MIN(<of> numeric-values)`

numeric-values is either a list of numeric values (numeric variables or numbers), separated by commas, or a list of variables in the form BASE*n*–BASE*m* (e.g., X1–X5). If the latter style is used, you must place the word "of" before the list.

If all of the arguments are missing, this function will return a missing value.

Examples

For these examples, $X = 5$, $Y = .$, $Z = 7$, $X1 = 1$, $X2 = 2$, $X3 = 3$, $X4 = .$

Function	Returns
MIN(X,Y,Z)	5
MIN(OF X1-X4)	1
MIN(10,20,. ,30,40,50)	10
MIN(OF X1-X4,Y,Z)	1

See Program 7.5 for an example.

Function: MAX

Purpose: To determine the largest non-missing value in a list of numeric values.

Syntax: `MAX(<of> numeric-values)`

numeric-values is either a list of numeric values (numeric variables or numbers), separated by commas, or a list of variables in the form BASE*n*–BASE*m* (e.g., X1–X5). If the latter style is used, you must place the word "of" before the list.

If all of the arguments are missing, this function will return a missing value.

Examples

For these examples, X = 5, Y = ., Z = 7, X1 = 1, X2 = 2, X3 = 3, X4 = .

Function	Returns
MAX(X,Y,Z)	7
MAX(OF X1-X4)	3
MAX(10,20,. ,30,40,50)	50
MAX(OF X1-X4,Y,Z)	7

Program 7.5: Program to read hourly temperatures and determine the daily minimum and maximum temperature

```
***Primary functions: MIN and MAX;

data min_max_temp;
   informat Date mmddyy10.;
   input Date;
   input Temp1-Temp24;
   Min_temp = min(of Temp1-Temp24);
   Max_temp = max(of Temp1-Temp24);
   keep Min_temp Max_temp Date;
   format Date mmddyy10.;
datalines;
05/1/2002
38 38 39 40 41 42 55 58 60 60 59 62 66 70 75 77 60 59 58 57 54 52 51 50
05/02/2002
36 41 39 40 41 46 57 59 63 . 59 62 64 72 79 80 78 62 62 62 60 50 55 55
;
```

```
title "Listing of Data Set MIN_MAX_TEMP";
proc print data=min_max_temp noobs;
run;
```

Explanation

Here, you have the temperature for each hour of the day and want to determine the minimum and maximum values for the day. Note that the one missing value for the second date does not result in a missing value for the minimum or maximum value. Here is the output listing:

```
Listing of Data Set MIN_MAX_TEMP

      Date    Min_temp    Max_temp

05/01/2002       38          77
05/02/2002       36          80
```

A Word about the SMALLEST and LARGEST Functions

These functions are somewhat different from the other functions described in this chapter. Given a list of variables, these functions can return the value of the nth largest or nth smallest value in the list. These functions can be thought of as generalized versions of MIN and MAX. Both of these functions ignore missing values in their calculations (as do MIN and MAX).

Function: SMALLEST

Purpose: To determine the value of the nth smallest value in a list of numeric values. This ordering ignores missing values.

Syntax: SMALLEST(N, <of> *numeric-values*)

N is a number from 1 to the maximum number of numeric values in the list. If N is larger than the number of values, the function returns a missing value and an error is written to the SAS log. If N is smaller than the number of values but greater than the number of non-missing values, the function returns a missing value, but no error is written to the SAS log.

numeric-values is either a list of numeric values (numeric variables or numbers), separated by commas, or a list of variables in the form BASE*n*–

BASE*m* (e.g., X1–X5). If the latter style is used, you must place the word "of" before the list.

Note: SMALLEST(1, *numeric-values*) is equivalent to MIN(*numeric-values*).

Examples

For these examples, X = 5, Y = ., Z = 7, X1 = 1, X2 = 2, X3 = 3, X4 = .

Function	Returns
SMALLEST(1,X,Y,Z)	5
SMALLEST(2, X,Y,Z)	7
SMALLEST(3, X,Y,Z)	. (missing)
SMALLEST(3, OF X1-X4)	3
SMALLEST(1, 10,20,. ,30,40,50)	10

Program 7.6: Computing the three lowest golf scores for each player (using the SMALLEST function)

```
***Primary function: SMALLEST;

data golf;
   infile datalines missover;
   input ID $ Score1-Score8;
   Lowest = smallest(1 ,of Score1-Score8);
   Next_lowest = smallest(2, of Score1-Score8);
   Third_lowest = smallest(3, of Score1-Score8);
datalines;
001 100 98 . . 96 93
002 90 05 07 99 103 106 110
003 110 120
;
title "Listing of Data Set GOLF";
proc print data=golf noobs heading=h;
run;
```

Explanation

First, a brief word on the INFILE statement. Since the data values follow a DATALINES statement, the special fileref DATALINES is used in an INFILE statement so that you can use the MISSOVER option. That option is needed because some of the lines of data contain fewer than eight scores.

Next, the SMALLEST function is used to extract the three lowest non-missing scores. The listing follows:

```
Listing of Data Set GOLF


ID      Score1     Score2     Score3     Score4     Score5     Score6

001      100         98         .          .          96         93
002       90          5         7         99         103        106
003      110        120         .          .          .          .

                               Next_      Third_
Score7    Score8    Lowest     lowest     lowest

  .         .          93         96         98
 110        .           5          7         90
  .         .         110        120          .
```

Function: LARGEST

Purpose: To determine the value of the nth highest value in a list of numeric values. This ordering ignores missing values.

Syntax: `LARGEST(N, <of> numeric-values)`

N is a number from 1 to the maximum number of numeric values in the list. If N is larger than the number of values, the function returns a missing value and an error is written to the SAS log. If N is smaller than the number of values but greater than the number of non-missing values, the function returns a missing value, but no error is written to the SAS log.

numeric-values is either a list of numeric values (numeric variables or numbers), separated by commas, or a list of variables in the form BASE*n*–BASE*m* (e.g., X1–X5). If the latter style is used, you must place the word "of" before the list.

Note: `LARGEST(1, numeric-values)` is equivalent to `MAX(numeric-values)`.

Examples

For these examples, X = 5, Y = ., Z = 7, X1 = 1, X2 = 2,
X3 = 3, X4 = .

Function	Returns
LARGEST(1,X,Y,Z)	7
LARGEST(2, X,Y,Z)	5
LARGEST(3, X,Y,Z)	. (missing)
LARGEST(3, OF X1-X4)	1
LARGEST(1, 10,20,. ,30,40,50)	50

Program 7.7: Computing a grade based on the five highest scores

```
***Primary function: LARGEST, N;

***This program will compute a grade based on the 5 highest
   (out of 9) scores.  If there are fewer than 5 non-missing scores,
   a missing value will be returned;

data high_5;
   input Score1-Score9;
   array Score[9];
   if n(of Score1-Score9) lt 5 then return;
   Sum = 0;
   do i = 1 to 5;
      Sum = Sum + largest(i,of Score1-Score9);
   end;
   Grade = Sum / 5;
   drop i;
datalines;
90 100 89 88 10 . . 29 77
. . . . . 100 99 98 97
10 20 30 40 50 60 70 80 90
;
title "Listing of Data Set HIGH_5";
proc print data=high_5 noobs;
run;
```

Explanation

The LARGEST function is used here to select the five highest scores. In case there are fewer than five non-missing scores, GRADE will be missing. The following listing of HIGH_5 shows that this program worked as desired:

\multicolumn{11}{c}{Listing of Data Set HIGH_5}										
Score1	Score2	Score3	Score4	Score5	Score6	Score7	Score8	Score9	Sum	Grade
90	100	89	88	10	.	.	29	77	444	88.8
.	.	.	.	.	100	99	98	97	.	.
10	20	30	40	50	60	70	80	90	350	70.

A Macro to Average Test Scores Where One or More of the Lowest Scores Are Dropped

For any teachers reading this book, I have included a macro of my own to average any number of test scores where you drop the lowest grade or grades from the calculation. This program does not give students zeros when they take fewer than the minimum number of quizzes. Rather, it computes the average of the actual number of quizzes taken. You may want to modify this program to penalize students taking fewer than the minimum number of quizzes. I think that the practice of dropping lowest grades helps keep students motivated if they score poorly on one or two tests.

Please note that with SAS 9.2, the CALL SORTN routine makes this problem even simpler (see Program 3.1).

Program 7.8: Macro to compute an average of *n* scores where the lowest *m* scores are dropped

```
***Primary functions: N, MIN, LARGEST;
*------------------------------------------------------------------*
| Macro: DROP_N                                                    |
| Purpose: Takes the average of "n" scores by dropping the         |
|          lowest "m" scores from the calculation.                 |
| Arguments: Dsn      Data set name                                |
|            Base     Base of variable name holding scores         |
|            N        Total number of scores                       |
|            N_drop   Number of scores to drop                     |
|            Varname  The name of the variable to hold the         |
|                     average                                      |
|            Note: The average variable is added to the original   |
|                  data set.                                       |
| Example: %macro drop_n(Roster=,                                  |
|                        Quiz=,                                    |
|                        N=12,                                     |
|                        N_drop=2,                                 |
|                        Varname=Average);                         |
*------------------------------------------------------------------*;
```

```
%macro drop_n(Dsn=, Base=, N= , N_drop=, Varname=);
   data &Dsn;
      set &Dsn;
      N_of_scores = n(of &Base.1-&Base&n);
      Min = min(N_of_scores,%eval(&N - &N_drop));
      Sum = 0;
      do i = 1 to Min;
         Sum + largest(i,of &Base.1-&Base&n);
      end;
      &Varname = Sum / Min;
      ***replace min with %eval(&N - &N_drop) to penalize
         students taking fewer than the minimum number of quizzes;
      drop i N_of_scores Sum Min;
   run;
%mend drop_n;
```

Explanation

As with other macros in this book, only a brief explanation will be provided. The N function is used to determine the number of non-missing quiz scores. The MIN function finds the number of quizzes to include in the averaging process: either the total number of quizzes minus the number to be dropped or the number of non-missing quiz scores, whichever is smaller. The first argument in the LARGEST function is the nth largest value. Suppose there were 9 quizzes, and you want to drop the two lowest scores. If the student took all 9, the minimum of 9 and 7 is 7, and the DO loop adds up the 7 highest scores. If the student took 8 quizzes, the DO loop also adds up the 7 highest (thus only one score being dropped). Finally, for any student taking 7 or fewer quizzes, the average is simply the average of all quizzes taken.

Note: To give students zeros when they take fewer than the minimum number of quizzes, you just have to replace the denominator of the VARNAME statement with %EVAL(&N - &N_DROP) instead of MIN.

To test this macro, use the following short DATA step to create a SAS data set (ROSTER). The macro is then invoked, using the 9 quiz grades in the computation. In this example, the lowest 2 quiz grades are to be dropped.

```
data roster;
    input ID Quiz1-Quiz9;
datalines;
1 6 7 8 9 8 7 6 7 8
2 6 8 8 8 8 8 8 8 .
3 7 8 9 . . . . . .
4 9 9 1 9 9 9 9 . .
;
```

The calling sequence to compute a quiz average, dropping the lowest 2 scores is:

```
%drop_n(Dsn=roster, Base=Quiz, N=9, N_drop=2, Varname=Quiz_ave)
```

Finally, the PRINT procedure is used to list the contents of the ROSTER data set, now with the new variable QUIZ_AVE included.

```
title1 "Listing of Data Set QUIZ";
title2 "After running macro DROP_N";
proc print data=roster noobs;
run;
```

```
Listing of Data Set QUIZ
After running macro DROP_N


ID Quiz1 Quiz2 Quiz3 Quiz4 Quiz5 Quiz6 Quiz7 Quiz8 Quiz9 Quiz_ave

 1   6     7     8     9     8     7     6     7     8    7.71429
 2   6     8     8     8     8     8     8     8     .    8.00000
 3   7     8     9     .     .     .     .     .     .    8.00000
 4   9     9     1     9     9     9     9     .     .    7.85714
```

Function: PCTL

Purpose: To determine the value, in a list of variables, corresponding to a given percentile. For example, at the 25th percentile, one quarter of the scores would be below this value. This ordering ignores missing values. The PCTLn function uses the standard five SAS definitions for computing percentiles. The default (PCTL) is definition five. In practice, the differences in the definitions are more apparent when there are fewer values in the list and if there are tied values.

Syntax: PCTL<*n*>(*percentile, <of> numeric-values*)

n is the percentile definition (1 to 5). If *n* is omitted, definition five is used.

percentile is a number from 0 to 100, specifying the percentile to be returned.

numeric-values is either a list of numeric values (numeric variables or numbers), separated by commas or a list of variables in the form BASE*n*–BASE*m* (e.g., X1–X5). If the latter style is used, you must place the word "of" before the list.

Examples

For these examples, X1=10, X2=12, X3=15, X4=17, X5=20, and X6=.

Function	Returns
PCTL(20, OF X1-X6)	11
PCTL2(20, OF X1-X6)	10
PCTL(75, 3,5,4,6,5,7,6,8,7,8,.,.)	7

Program 7.9: Using the PCTL function to determine 25th, 50th, and 75th percentile in a list of values

```
***Primary function: PCTL
***Other functions: RAND;

***Generate data set for testing;
data temperature;
   array T[24]; ***temperature for each hour;
   do Day = 1 to 5;
      ***T values normally distributed with mean = 70 and
         standard deviation = 10;
      do Hour = 1 to 24;
         T[hour] = rand('normal',70,10);
      end;
      output;
   end;
   keep T1-T24 Day;
run;
```

```
data percentile;
   set temperature;
   P25 = pctl(25, of T1-T24);
   P50 = pctl(50, of T1-T24);
   P75 = pctl(75, of T1-T24);
   IQR = P75 - P25;
   label P25 = "25th Percentile"
         P50 = "Median"
         P75 = "75th Percentile"
         IQR = "Inter-quartile range";
run;

title "Listing of Data Set PERCENTILE";
proc print data=percentile noobs label;
   id Day;
   var P25 P50 P75 IQR;
run;
```

Explanation

The first DATA step creates a test data set of 24 temperatures for each of five days. The next DATA step computes the 25th, 50th (median), and 75th percentile, as well as the difference between the 25th and 75th percentile, called the interquartile range (IQR).

```
Listing of Data Set PERCENTILE

                                                Inter-
            25th                      75th     quartile
Day      Percentile    Median     Percentile    range

1          66.2169     73.6732      80.1367     13.9198
2          64.8195     72.4530      76.5298     11.7103
3          65.7105     70.0485      75.2977      9.5872
4          63.1399     69.7068      75.5965     12.4565
5          65.1751     68.2329      79.8730     14.6979
```

Function: ORDINAL

This function is somewhat different from the functions described previously in this chapter. Given a list of numeric values, this function can return the value of the *n*th lowest value. Unlike the MIN, MAX, LARGEST, and SMALLEST functions, the ORDINAL function **includes** missing values in its calculations.

Purpose: To determine the value of the *n*th lowest value in a list of numeric values. This ordering includes missing values.

Syntax: ORDINAL(*ordinality*, <of> *numeric-values*)

ordinality is a number from 1 to the maximum number of numeric values in the list.

numeric-values is either a list of numeric values (numeric variables or numbers), separated by commas, or a list of variables in the form BASE*n*–BASE*m* (e.g., X1–X5). If the latter style is used, you must place the word "of" before the list.

Examples

For these examples, X = 5, Y = ., Z = 7, X1 = 1, X2 = 2, X3 = 3, X4 = .

Function	Returns
ORDINAL(1,X,Y,Z)	. (missing value)
ORDINAL(2, X,Y,Z)	5
ORDINAL(3, X,Y,Z)	7
ORDINAL(3, OF X1-X4)	2
ORDINAL(1, 10,20,. ,30,40,50)	. (missing value)

Program 7.10: Program to compute a quiz grade by dropping none, one, or two of the lowest quiz scores, depending on how many quizzes were taken

```
***Primary function: ORDINAL
***Other functions: N, MEAN, and SUM;

data quiz_ave;
   input ID $ Quiz1-Quiz9;
   N_of_quizzes = n(of Quiz1-Quiz9);
   if n_of_quizzes  = 9 then
      Quiz_ave = (sum(of Quiz1-Quiz9) -
                  ordinal(1,of Quiz1-Quiz9) -
                  ordinal(2,of Quiz1-Quiz9))/7;
   else if N_of_quizzes = 8 then
      Quiz_ave = (sum(of Quiz1-Quiz9) - ordinal(2,of Quiz1-Quiz9))/7;
   else Quiz_ave = mean(of Quiz1-Quiz9);
```

```
datalines;
001 6 7 8 9 8 7 6 7 8
002 6 8 8 8 8 8 8 8 .
003 7 8 9 . . . . . .
004 9 9 1 9 9 9 9 . .
;
title "Listing of Data Set QUIZ_AVE";
proc print data=quiz_ave noobs heading=h;
run;
```

Explanation

Before we discuss this program, I should point out that the CALL SORTN routine makes this a very simple problem (see Program 3.1). In addition, the LARGEST function would also make for a much simpler solution. However, I have decided to include this more complicated program as a demonstration of how to use the ORDINAL function.

This example was developed to compute grades for a course that I taught at the University of Medicine and Dentistry School of Public Health in New Jersey. Students in my biostatistics course took weekly quizzes. If they took all the quizzes, I dropped the two lowest scores. If they missed one quiz, they got to drop the lowest score. If two or more quizzes were missed, no scores were dropped. Using the ORDINAL function is a great help with this problem. The term ORDINAL(1, OF QUIZ1-QUIZ9) returns the lowest (or missing) quiz score and the statement ORDINAL(2, OF QUIZ1-QUIZ9) returns the second lowest quiz score (including missing values).

Therefore, if all 9 quizzes were taken, the quiz average is the total of all 9 scores, minus the two lowest scores, divided by 7. If only 8 quizzes were taken, you subtract the second lowest quiz score (remember, the value of ORDINAL(1, OF QUIZ1-QUIZ9) will be a missing value in this situation) from the total and divide by 7. Finally, if the number of non-missing quiz scores is not 8 or 9, you compute the mean of the non-missing quiz scores using the MEAN function. You can inspect the following listing to see that this program works as desired:

```
Listing of Data Set QUIZ_AVE


ID     Quiz1    Quiz2    Quiz3    Quiz4    Quiz5    Quiz6

001      6        7        8        9        8        7
002      6        8        8        8        8        8
003      7        8        9        .        .        .
004      9        9        1        9        9        9


                          N_of_
Quiz7    Quiz8    Quiz9    quizzes    Quiz_ave

  6        7        8         9        7.71429
  8        8        .         8        8.00000
  .        .        .         3        8.00000
  9        .        .         7        7.85714
```

An interesting alternative to Program 7.10 is shown next. Here, a DO loop is used to add up the quiz scores, starting from the highest to the lowest. Look at the program and then study the following explanation:

Program 7.11: Alternate version of Program 7.10

```
***Primary function: ORDINAL
***Other functions: N and MIN;

data quiz_ave2;
   input ID $ Quiz1-Quiz9;
   N_of_quizzes = n(of Quiz1-Quiz9);
   Sum = 0;
   do i = 1 to min(N_of_quizzes,7);
      Sum + ordinal(10 - i,of Quiz1-Quiz9);
   end;
   Quiz_ave = Sum / min(N_of_quizzes,7);
datalines;
001 6 7 8 9 8 7 6 7 8
002 6 8 8 8 8 8 8 8 .
003 7 8 9 . . . . . .
004 9 9 1 9 9 9 9 . .
;
title "Listing of Data Set QUIZ_AVE2";
proc print data=quiz_ave2 noobs heading=h;
run;
```

Explanation

Although this program is shorter than the previous program, it's actually a bit more complicated. However, it is a good way to demonstrate the N, MIN, and ORDINAL functions.

The number of quizzes taken is computed using the N function and is assigned to the variable N_OF_QUIZZES. For each iteration of the DATA step, the SUM is initialized to 0. The DO loop starts from 1 and ends at the number of quizzes taken or the number 7, whichever is smaller. Next, each quiz score is added to the sum using a SUM statement, starting from the highest quiz score to the lowest, non-missing quiz score. Finally, the average is computed by dividing this sum by the number of quizzes taken or the number 7, whichever is smaller. The output from this program is identical to the output from Program 7.10 and will not be shown.

Using the ORDINAL Function to Sort Values within an Observation

As of SAS 9.2, the two CALL routines (CALL SORTN and CALL SORTC) provide a way to sort values within an observation (see Chapter 3). Therefore, it is no longer necessary to write a program to accomplish this task. The program was left here as a good demonstration of the ORDINAL function.

The program and macro in this section use the ORDINAL function to sort numerical values within an observation. For example, if you have variables X1–X10 and want to rearrange the values so that X1 holds the lowest value, X2, the next lowest value, and so forth, you can use this program to accomplish the task.

Program 7.12: Sorting values within an observation

```
***Primary function: ORDINAL
***Other functions: CALL SORTN;

data sort;
   input X1-X10;
   array X[10] X1-X10;
   array Sort_x[10] Sort_x1-Sort_x10;
   do i = 1 to 10;
      Sort_x[i] = ordinal(i,of X1-X10);
   end;
   drop i;
```

```
datalines;
5 2 9 1 3 6 . 22 7 0
;
title "Listing of Data Set SORT";
proc print data=sort noobs heading=h;
run;

*Using call SORTN;
data sort;
   input X1-X10;
   call sortn(of X1-X10);
datalines;
5 2 9 1 3 6 . 22 7 0
;
title "Listing of Data Set SORT";
proc print data=sort noobs heading=h;
run;
```

Explanation

In one of my earlier books, I wrote a fairly long and complicated program to sort values within an observation, using a bubble sort algorithm. That was before I became familiar with the ORDINAL function. The use of this function greatly simplifies this task. Program 7.12 leaves the original values unchanged and creates a new set of variables to hold the sorted values. The key to the whole program is to use the DO loop index (I) as the first argument in the ORDINAL function, thus resulting in the new set of variables (SORT_Xn) in sorted order.

The output from the first PROC PRINT is shown next (the listing from the second PROC PRINT is identical):

```
Listing of Data Set SORT

X1 X2 X3 X4 X5 X6 X7 X8 X9 X10 Sort_x1 Sort_x2 Sort_x3 Sort_x4

 5  2  9  1  3  6  . 22  7  0      .       0       1       2

Sort_x5    Sort_x6    Sort_x7    Sort_x8    Sort_x9    Sort_x10

   3          5          6          7          9          22
```

Once again, please remember that with SAS 9.2, CALL SORTN will sort values in a list of variables. This program was left in as a demonstration of the ORDINAL function and a simple macro.

The sort macro presented in the following program will sort the values for a set of variables and return the sorted values to the original variable names. For example, if your original variables were X1–X10, the same ten variables would have the values in sorted order after the macro was run. Here is the program:

Program 7.13: Macro to sort values within an observation

```
***Primary function: ORDINAL;

*Note: CALL SORTN can be used to sort values within and observation.
 This program is included to provide another example of the ORDINAL
 function as well as temporary arrays;

%macro sort_array(Dsn=,    /* Data set name                    */
                  Base=,   /* Prefix of variable name
                              eg. if variables are Ques1-Ques10,
                              Base would be Ques                */
                  N_of_elements= /* Number of variables        */);
   data &Dsn;
      set &Dsn;
      array &Base[&n_of_elements];
      array Temp[&N_of_elements] _temporary_;
      ***transfer values to temporary arry;
      do i = 1 to &N_of_elements;
         Temp[i] = ordinal(i,of &Base.1 - &Base&N_of_elements);
      end;
      ***put them back in the original variable names in order;
      do i = 1 to &N_of_elements;
         &Base[i] = Temp[i];
      end;
      drop i;
   run;
%mend  sort_array;
```

Explanation

This program uses a temporary array as a place to hold the sorted values so that they can be placed back into the original variable names. You could easily modify this program if you want to preserve the original variables and create a new series of variables to hold the sorted values. The following code shows the data set before the macro is called, the macro call, and a listing of the data set after the sorting has been done.

```
title "Listing of Data Set SORT before macro";
proc print data=sort noobs;
   var x1-x10;
run;

%sort_array(Dsn=sort, Base=x, N_of_elements=10)

title "Listing of Data Set SORT after macro";
proc print data=sort noobs;
   var x1-x10;
run;
```

The two listings below show that the macro is working as desired.

```
Listing of Data Set SORT before macro

X1    X2    X3    X4    X5    X6    X7    X8    X9    X10

 5     2     9     1     3     6     .    22     7     0

Listing of Data Set SORT after macro

X1    X2    X3    X4    X5    X6    X7    X8    X9    X10

 .     0     1     2     3     5     6     7     9     22
```

Using the STD Function to Perform a *t*-Test

This last section uses the STD function, along with several others, to perform a *t*-test (a statistical test to compare the means of two groups) where all the values for one group are in one observation and all the values for the other group are in the next observation. In order to use PROC TTEST, a new data set would have to be created where each value was in a separate observation, along with the GROUP variable.

Program 7.14: Performing a *t*-test where values for each group are in a single observation

```
***Primary function: STD
***Other functions: N, MEAN, SQRT, PROBT, ABS;

title "T-Test Calculation";
data _null_;
   file print;
   infile datalines missover;
   input Group $ x1-x50;
   retain n1 Mean1 SD1 Group1;
   if _n_ = 1 then do;
      Group1 = Group;
      N1 = n(of x1-x50);
      Mean1 = mean(of x1-x50);
      SD1 = std(of x1-x50);
   end;
   if _n_ = 2 then do;
      Group2 = Group;
      N2 = n(of x1-x50);
      Mean2 = mean(of x1-x50);
      SD2 = std(of x1-x50);
      Diff = Mean1 - Mean2;
      SD_pooled_2 = ((n1-1)*SD1**2 + (n2-1)*SD2**2)/(n1 + n2 - 2);
      T = abs(Diff) / sqrt(SD_pooled_2/n1 + SD_pooled_2/n2);
      Prob = 2*(1 - probt(T,n1+n2-2));

      ***Prepare the report;
      put @1   "Group " Group1 +(-1) ":"
          @10 "N = " n1 3.
          @20 "Mean = " Mean1 7.3
          @35 "SD = " SD1 7.4 /
          @1   "Group " Group2 +(-1) ":"
          @10 "N = " n2   3.
          @20 "Mean = " Mean2 7.3
          @35 "SD = " SD2 7.4 /
          @1 "Difference = " Diff 7.3
          @25 "T = " T 7.4
          @40 "P(2-tail) = " Prob 7.4;
   end;
datalines;
A 4 5 8 7 6 5 7
B 9 7 8 8 6 7 9 9 11
;
```

Explanation

Since each group of values resides within a single observation, the SAS automatic variable _N_ allows you to determine if you are reading the first or second line of data. The three descriptive statistics functions N, MEAN, and STD compute the number of non-missing values, the mean, and the standard deviation, respectively. After these values have been computed for the second group, the steps to compute a *t*-value are performed. The SQRT function takes the square root of its argument. The ABS function takes the absolute value of the difference so that our *t*-value will always be positive. Finally, in order to print out the two-tailed *p*-value (the famous *p*-value that all researchers hope is less than .05), the PROBT function is used. This function computes the cumulative probability under the *t* distribution. Here is the output from this program:

```
                         T-Test Calculation
Group A: N =    7    Mean =    6.000 SD =   1.4142
Group B: N =    9    Mean =    8.222 SD =   1.4814
Difference =   -2.222    T =   3.0349    P(2-tail) =   0.0089
```

Chapter 8

Mathematical and Numeric Functions

Introduction

Most of the functions in this chapter perform basic mathematical tasks such as taking the log or exponent of a value. The MOD function is an exception. Although it is used in mathematical expressions, it has other, often overlooked, uses that will be covered here.

Commonly Used Mathematical Functions

Function: CONSTANT

Purpose: To compute numerical constants: (pi, e, and Euler's constant) and to determine various machine constants (the largest integer stored in *n* bytes, the largest double-precision number, the smallest double-precision number (its log and square root), the machine precision constant). Also, the log and square root of the last three constants.

Syntax: CONSTANT('*constant*' <,*parm*>)

constant, placed in quotation marks, is one of the following:

Mathematical Constants

e	is the base of natural logarithms.
EULER	is Euler's constant.
PI	is pi.

Machine Constants

EXACTINT	is the largest integer stored in *n* bytes (specified by *parm* = 2 to 8).
BIG	is the largest double-precision number stored in 8 bytes on your computer.
LOGBIG	is the logarithm of BIG. The default base is E, but you can specify a base with *parm*.
SQRTBIG	is the square root of BIG.
SMALL	is the smallest double-precision number stored in 8 bytes on your computer.
LOGSMALL	is the logarithm of SMALL (base specified with *parm*).
SQRTSMALL	is the square root of SMALL.

Machine Precision

MACEPS	is a machine-precision constant. This number is used to determine if one number is larger than another.
LOGMACEPS	is the log of MACEPS (base specified with *parm*).
SQRTMACEPS	is the square root of MACEPS.

Examples

These examples were run on a Windows platform.

Function	Returns
CONSTANT('PI')	3.1415926536
CONSTANT('EXACTINT',3)	8192
CONSTANT('EXACTINT',4)	2097152
CONSTANT('EXACTINT',8)	9.0071993E15
CONSTANT('BIG')	1.797693E308
CONSTANT('SMALL')	2.22507E-308
CONSTANT('MACEPS')	2.220446E-16

Program 8.1: Determining mathematical constants and machine constants using the CONSTANT function

```
***Primary function: CONSTANT;

title;
data _null_;
   file print;
   Pi = constant('pi');
   e = constant('e');

   put "Mathematical Constants" /
       "Pi = " Pi /
       "e = " e //
       "Largest Integers stored in 'n' bytes:";
   do Bytes = 3 to 8;
      Int = constant('exactint',Bytes);
      put +5 "Largest integer stored in " Bytes "bytes is: " Int;
   end;
   Large = constant('big');
   Small = constant('small');
   Precision = constant('maceps');
   put / "Machine Constants" /
         "Largest 8 Byte Value is: " Large /
         "Smallest 8 Byte Value is: " Small /
         "Precision constant is: " Precision;
run;
```

Explanation

The CONSTANT function is used here to list some of the mathematical and machine constants. If you plan to change the default length of 8 for numerical values, it would be an excellent idea to use the CONSTANT function to determine the largest integer you could store exactly in the number of bytes you choose. Here is the listing from this program (run on a Windows platform):

```
Mathematical Constants
Pi = 3.1415926536
e = 2.7182818285

Largest Integers stored in 'n' bytes:
    Largest integer stored in 3 bytes is: 8192
    Largest integer stored in 4 bytes is: 2097152
    Largest integer stored in 5 bytes is: 536870912
    Largest integer stored in 6 bytes is: 137438953472
    Largest integer stored in 7 bytes is: 3.5184372E13
    Largest integer stored in 8 bytes is: 9.0071993E15

Machine Constants
Largest 8 Byte Value is: 1.797693E308
Smallest 8 Byte Value is: 2.22507E-308
Precision constant is: 2.220446E-16
```

Function: MOD

You probably have to think back to ninth grade to remember what modular arithmetic is. The MOD function returns the remainder of one number when it is divided by another. For example, 5 mod 3 is 2; 15 mod 3 is 0, 16 mod 3 is 1, and so forth. Two applications of the MOD function are presented here: one to select every nth observation from a SAS data set, and the other, a programming trick to use the MOD function as a toggle switch, allowing you to alternate two activities.

Purpose: To perform modular arithmetic. Also useful to select or process every *n*th observation.

Syntax: MOD *(numeric-value, modulo)*

numeric-value is a numeric variable or expression.

modulo is a constant or a numeric variable. The function returns the remainder after *numeric-value* is divided by *modulo*.

Examples

Function	Returns
MOD(15,4)	3
MOD(4,15)	4
MOD(7,2)	1

Program 8.2: Using the MOD function to choose every *n*th observation from a SAS data set

```
***Primary function: MOD
***Other functions: INT, RANUNI;

***Create test input data set;
data big;
   do Subj = 1 to 20;
      x = int(10*ranuni(0)); /* Random integers from 0 to 9 */
      output;
   end;
run;

%let n = 4; /* Every 4th observation will be selected */

data every_n;
   set big;
   if mod(_n_ ,&n) = 1;
   /* Selects every nth observation, starting with the 1st */
run;

title "Listing of data set EVERY_N";
proc print data=every_n noobs;
run;
```

Explanation

The first DATA step (the data set BIG) is included to produce a test data set with 20 subjects. The key to choosing every *n*th observation is the subsetting IF statement, using the MOD function to provide the condition. When the value of the automatic variable _N_ is 1, 5, 9, etc., the condition will be true and the observation is added to data set EVERY_N. The following short listing demonstrates that this program works as advertised.

```
Listing of data set EVERY_N

Subj    x

  1     5
  5     7
  9     8
 13     8
 17     1
```

Program 8.3: Using the MOD function as a toggle switch, alternating group assignments

```
***Primary function: MOD;
***In this program, we want to assign every other subject into
   group A or group B;

data switch;
   do Subj = 1 to 10;
      if mod(Subj,2) eq 1 then Group = 'A';
      else Group = 'B';
      output;
   end;
run;
title "Listing of data set SWITCH";
proc print data=switch noobs;
run;
```

Explanation

Since the second argument of the MOD function is a 2, the function will return either 1's or 0's alternately. This is a useful feature that can be used in any program where you need to perform two different activities in an alternating fashion. Here is the listing from PROC PRINT:

```
Listing of data set SWITCH

Subj    Group

  1       A
  2       B
  3       A
  4       B
  5       A
  6       B
  7       A
  8       B
  9       A
 10       B
```

Function: ABS

Purpose: To take the absolute value of its argument (i.e., throw away the minus sign if there is one).

Syntax: ABS *(numeric-value)*

numeric-value is a number, a numeric variable, or expression.

Examples

Function	Returns
ABS (-15)	15
ABS (8)	8

Program 8.4: Computing chi-square with Yates' correction, which requires an absolute value

```
***Primary function: ABS
***Other function: PROBCHI;

data yates;
   input A B C D;
   N = A + B + C + D;
   Yates = ( abs(A*D - B*C) - N/2)**2 * N /
           ( (A+B)*(C+D)*(A+C)*(B+D) );
      Prob_Yates = 1 - probchi(Yates,1);
datalines;
10 20 30 40
2 9 8 5
;
title "listing of data set YATES";
proc print data=yates noobs;
run;
```

Explanation

The formula for chi-square with Yates' correction takes the absolute value of the numerator, thus requiring the use of the ABS function. Here is the listing:

```
listing of data set YATES


                                   Prob_
 A     B    C    D    N     Yates   Yates

10    20   30   40   100   0.44643  0.50404
 2     9    8    5    24   2.99700  0.08342
```

Function: SQRT

Purpose: To take the square root of a value.

Syntax: SQRT *(numeric-value)*

numeric-value is a numeric variable or expression greater than or equal to zero.

Examples

Function	Returns
SQRT(16)	4
SQRT(-4)	missing value

Program 8.5:   Program to compute and print out a table of integers and square roots

```
***Primary function: SQRT;

options ps=15; /* So the panels will display in PROC REPORT */
data square_root;
   do n = 1 to 40;
      Square_root = sqrt(n);
      output;
   end;
run;

title "Table of Integers and Square Roots";
proc report data=square_root nowd panels=99;
   columns n Square_root;
   define n / display width=3 format=3.0;
   define Square_root /'Square Root' width=7 format=7.6;
run;
```

Explanation

This very straightforward program uses a DO loop to generate the integers from 1 to 40. The SQRT function computes the square root of each integer, and the OUTPUT statement writes an observation to the SAS data set. In this example, PROC REPORT was used so that the PANELS option could be used. This option creates telephone-book style output. If the number of panels is more than will fit on the page, PROC REPORT attempts to fit as many panels as possible. Finally, in the following listing, the page size option was set to 15 so that the panels would appear in the space allowed.

```
Table of Integers and Square Roots

      Square           Square           Square           Square
  n    Root       n     Root       n     Root       n     Root
  1   1.00000     12   3.46410     23   4.79583     34   5.83095
  2   1.41421     13   3.60555     24   4.89898     35   5.91608
  3   1.73205     14   3.74166     25   5.00000     36   6.00000
  4   2.00000     15   3.87298     26   5.09902     37   6.08276
  5   2.23607     16   4.00000     27   5.19615     38   6.16441
  6   2.44949     17   4.12311     28   5.29150     39   6.24500
  7   2.64575     18   4.24264     29   5.38516     40   6.32456
  8   2.82843     19   4.35890     30   5.47723
  9   3.00000     20   4.47214     31   5.56776
 10   3.16228     21   4.58258     32   5.65685
 11   3.31662     22   4.69042     33   5.74456
```

Functions That Work with Exponentiation and Logarithms

These functions have been grouped together since they are all related and have similar syntax. I will describe the three functions and then follow with a single program that uses all three.

Function: EXP

Purpose: To return the value of the exponential function. EXP computes the value of e (the base of natural logarithms), raised to the power of the argument.

Syntax: EXP(*numeric-value*)

numeric-value is a number, numeric variable, or expression.

Examples

Function	Returns
EXP(2.718)	15.150
EXP(100)	2.688
EXP(0)	1

Function: **LOG**

Purpose: To compute the natural logarithm (base e) of an argument.

Syntax: LOG(*numeric-value*)

numeric-value is a positive numeric variable or expression.

Examples

Function	Returns
LOG(2.718)	1.000
LOG(100)	4.605
LOG(1000)	6.908

Function: **LOG10**

Purpose: To compute the base-10 logarithm of an argument.

Syntax: LOG10(*numeric-value*)

numeric-value is a positive numeric variable or expression.

Examples

Function	Returns
LOG10(2.718)	.434
LOG10(100)	2
LOG10(1000)	3

Program 8.6: Creating tables of integers, their base-10 and base e logarithms and their value taken to the *n*th power

```
***Primary functions: EXP, LOG, LOG10;
***Program to print a pretty table of integers (from 1 to 100),
   results of the exponential function, base ten and base E logs;

data table;
   do n = 1 to 40;
      e = exp(n);
      ln = log(n);
      log = log10(n);
      output;
   end;
run;

title "Table of Exponents, Natural and Base 10 Logs";
proc report data=table nowd panels=99;
   columns n e ln log;
   define n / display right width=3 format=3.;
   define e / display right width=8 format=best8. 'Exp';
   define ln / display right width=7 format=7.4 'Natural Log';
   define log / display right width=7 format=7.4 'Base 10 Log';
run;
```

```
Table of Exponents, Natural and Base 10 Logs

              Natural    Base                    Natural    Base
   n     Exp      Log   10 Log      n      Exp       Log   10 Log
   1  2.718282  0.0000  0.0000     12  162754.8   2.4849  1.0792
   2  7.389056  0.6931  0.3010     13  442413.4   2.5649  1.1139
   3  20.08554  1.0986  0.4771     14  1202604    2.6391  1.1461
   4  54.59815  1.3863  0.6021     15  3269017    2.7081  1.1761
   5  148.4132  1.6094  0.6990     16  8886111    2.7726  1.2041
   6  403.4288  1.7918  0.7782     17  24154953   2.8332  1.2304
   7  1096.633  1.9459  0.8451     18  65659969   2.8904  1.2553
   8  2980.958  2.0794  0.9031     19  1.7848E8   2.9444  1.2788
   9  8103.084  2.1972  0.9542     20  4.8517E8   2.9957  1.3010
  10  22026.47  2.3026  1.0000     21  1.3188E9   3.0445  1.3222
  11  59874.14  2.3979  1.0414     22  3.5849E9   3.0910  1.3424
Table of Exponents, Natural and Base 10 Logs

              Natural    Base                    Natural    Base
   n     Exp      Log   10 Log      n      Exp       Log   10 Log
  23  9.7448E9  3.1355  1.3617     34  5.835E14   3.5264  1.5315
  24  2.649E10  3.1781  1.3802     35  1.586E15   3.5553  1.5441
  25    7.2E10  3.2189  1.3979     36  4.311E15   3.5835  1.5563
  26  1.957E11  3.2581  1.4150     37  1.172E16   3.6109  1.5682
  27   5.32E11  3.2958  1.4314     38  3.186E16   3.6376  1.5798
  28  1.446E12  3.3322  1.4472     39  8.659E16   3.6636  1.5911
  29  3.931E12  3.3673  1.4624     40  2.354E17   3.6889  1.6021
  30  1.069E13  3.4012  1.4771
  31  2.905E13  3.4340  1.4914
  32  7.896E13  3.4657  1.5051
  33  2.146E14  3.4965  1.5185
```

Factorial and Gamma Functions

Prior to SAS 7, SAS did not include the FACT function. It was necessary to use the relationship that GAMMA(N + 1) was equal to N!.

Function: FACT

Purpose: To take the factorial of its argument. For example, 4! = FACT(4) = 4 x 3 x 2 x 1 = 24.

Syntax: FACT(*numeric-value*)

> *numeric-value* is a numeric variable or expression that resolves to an integer that is greater than or equal to 0.

Examples

Function	Returns
FACT(4)	24
FACT(0)	1
FACT(20)	2.432902E18

Program 8.7: Creating a table of integers and factorials

```
***Primary function: FACT;

data factorial;
   do n = 1 to 12;
      Factorial_n = fact(n);
      output;
   end;
   format Factorial_n comma12.;
run;

title "Listing of data set FACTORIAL";
proc print data=factorial noobs;
run;
```

```
Listing of data set FACTORIAL

n     Factorial_n

1               1
2               2
3               6
4              24
5             120
6             720
7           5,040
8          40,320
9         362,880
10      3,628,800
11     39,916,800
12    479,001,600
```

Function: GAMMA

Purpose: To compute the value of the Gamma function. For positive integers,
Gamma(X) = (X − 1)!.

Syntax: GAMMA(*numeric-value*)

numeric-value is a numeric variable or expression greater than or equal
to 0.

Examples

Function	Returns
GAMMA(4)	6
GAMMA(1)	1
GAMMA(4.5)	11.631728397

Program 8.8: Demonstrating the GAMMA function

```
***Primary function: GAMMA;

data table;
   do x = 1 to 5 by .05;
      Gamma = gamma(x);
      output;
   end;
run;

ods rtf file='c:\books\functions\gamma.rtf';
goptions device=jpeg;
symbol v=none i=sm;
title "Graph of Gamma Function from 1 to 5";
proc gplot data=table;
   plot Gamma * x;
run;
ods rtf close;
```

Explanation

Since the GAMMA function is not restricted to integers, this program computes the value of the GAMMA function from 1 to 5 with increments of .05. The GPLOT procedure is used to display this graph.

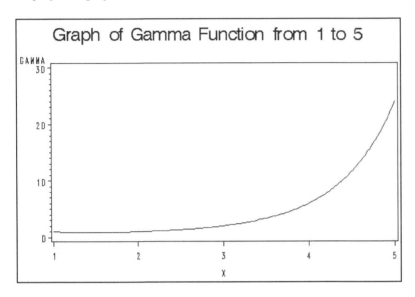

Miscellaneous Functions

Function: **IFN**

Purpose: To select a numeric value depending on whether a logical expression is true, false, or missing. (Please see Chapter 1, "Character Functions," for a description of the corresponding function IFC that deals with character data.)

Syntax: IFN *(logical-expression, if-true, if-false, <if-missing>)*

logical-expression is a numeric constant, value, or expression. Remember that any SAS numeric value that is not zero or missing is considered true.

Examples

For these examples, X=100, Y = . (missing).

Function	Returns
IFN(X gt 90,5,6)	5
IFN(X lt 90,5,6)	6
IFN(Y,1,2,3)	3

Program 8.9: Demonstrating the IFN Function

```
***Primary function: IFN;
***Other function: MISSING;
data bonus;
   input Name : $10. Sales;
   if not missing(Sales) then Bonus = ifn(Sales > 5000, 200, 100);
datalines;
Ron 6000
John 4000
Fred .
;
title "Listing of data set BONUS";
proc print data=bonus noobs;
run;
```

Explanation

When the value of SALES is greater than 5,000 (making the expression true), BONUS is set to 200. For non-missing SALES values less than 5,000, BONUS is equal to 100. Without the test for a missing value for SALES, the IFN function would return a BONUS amount of 100 for a missing SALES value, since the expression would be evaluated as false. The last (optional) argument in the IFN function is selected only when the value of the logical expression is missing.

Function: COALESCE

Purpose: To select the first non-missing value in a list of numeric arguments. This function is similar to the COALESCEC function that takes character arguments.

Syntax: COALESCE *(<of> Num-1 <,Num-2, ...>)*

Num-1, Num-n are numeric values or expressions. The COALESCE function returns the value of the first non-missing argument. If all the arguments have missing values, the function returns a missing value. If you use a list of variables in the form Var1–Varn, you need to precede the list with the word "of."

Examples

For these examples, X=100, Y = . (missing), Z = 300
Ques1 = ., Ques2 = 5, Ques3 = 8

Function	Returns
COALESCE(X,Y,Z)	100
COALESCE(Y,Z,X)	300
COALESCE(of Ques1-Ques3))	5

Program 8.10: Demonstrating the COALESCE function

```
***Primary function: COALESCE
data weigh_in;
   input Name $ Time1-Time3;
   Weight = coalesce(of Time1-Time3);
datalines;
John 180 . .
Fred . 250
Gary . . 350
Peter 200 . .
;
title "Listing of data set WEIGH_IN";
proc print data=weigh_in noobs;
run;
```

Explanation

In this example, each subject weights in at either time 1, time 2, or time 3. Rather than having to use logical expressions to test each value to find the first non-missing value, the COALESCE function make for a much more elegant solution. Here is the listing:

```
Listing of data set WEIGH_IN

Name     Time1    Time2    Time3    Weight

John      180       .        .        180
Fred       .       250       .        250
Peter     200       .        .        200
```

C h a p t e r   9

Random Number Functions

Introduction

This chapter focuses on the random number functions—functions that produce pseudo-random values according to a particular distribution. (Since computers can't create truly random numbers, the term pseudo-random numbers is the technically correct description of any computer-generated series of numbers. However, in this chapter, we will go ahead and call these **random numbers**.)

Only two of the numerous older random number functions are described here along with the new (SAS®9) RAND function that can generate random numbers that follow many of the well known probability distributions (such as uniform, normal, binomial, and Poisson). With the introduction of the RAND function, most of the older random number functions are not needed. The RAND function is also considered superior to the older functions, since it generates random numbers with a long period (where the sequence repeats) and with good statistical properties. If you would like to read a description of these older functions, please see Product Documentation in the Knowledge Base, available at http://support.sas.com/documentation.

Note: The wording of arguments in this book might differ from the wording of arguments in the Product Documentation.

Although most of the programs focus on statistical applications, you may find it useful to take a random subset of a SAS data set or to randomly generate test data sets, whether you are a statistician or a non-statistician.

A Word about Random Number Seeds

It is important to understand how SAS generates a series of random numbers using the random number functions and the CALL routines. There are some important differences.

All of the random number functions share a common series of seeds. The random number functions can either use a seed that you supply (in which case you get the same series of random numbers every time you run the same DATA step) or a 0 (or any negative value seed), in which case the first seed is generated using the computer clock. Regardless of what seed you designate for any subsequent uses of any random functions in your DATA step, it is only the **first** seed that matters in the random number functions.

If you choose to use one of the random number CALL routines, you can choose a new seed value any time you want. In addition, you can have a separate seed stream for each of several random number generators in your program.

To clarify how seed values affect a series of random numbers, look at the two following programs:

Program 9.1: Demonstrating differences between random functions and CALL routines (function example)

```
***Primary function: RANUNI;
data ran_function;
   do n = 1 to 3;
      x = ranuni(1234);
      y = ranuni(0);
      output;
   end;
run;

title "Listing of data set RAN_FUNCTION - First run";
proc print data=ran_function noobs;
run;
```

```
title "Listing of data set RAN_FUNCTION - Second run";
proc print data=ran_function noobs;
run;
```

Explanation

This program demonstrates how seed values affect the results of random number functions.
Notice in the following listing that when this program is run twice, you get the same series
of random numbers for both X and Y, even though you used a value of 0 for the seed of Y.

```
Listing of data set RAN_FUNCTION - First run

n     x           y

1    0.24381    0.089475
2    0.38319    0.097928
3    0.25758    0.088250

Listing of data set RAN_FUNCTION - Second run

n     x           y

1    0.24381    0.089475
2    0.38319    0.097928
3    0.25758    0.088250
```

**Program 9.2: Demonstrating differences between random functions and
CALL routines (CALL routine example)**

```
data ran_call;
   Seed1 = 1234;
   Seed2 = 0;
   do n = 1 to 3;
      call ranuni(Seed1,x);
      call ranuni(Seed2,y);
      output;
   end;
run;

title "Listing of data set RAN_CALL - First run";
proc print data=ran_call noobs;
run;

title "Listing of Data Set RAN_CALL - Second Run";
proc print data=ran_call noobs;
run;
```

Explanation

Notice that the series of random numbers for Y in this example is different each time the program runs. This demonstrates one advantage of using random number CALL routines instead of random number functions: you can have complete control over the seed values. If you look closely at the values for X in the following listing, you will notice that the three values of X in the first three observations correspond to the first three random numbers produced in Program 9.1 (X in observation 1; Y in observation 1; X in observation 2).

```
Listing of data set RAN_CALL - First run

  Seed1         Seed2       n      x         y

523580480     911504187     1   0.24381   0.42445
192146343    1784774646     2   0.08948   0.83110
822902235     166306989     3   0.38319   0.07744

Listing of Data Set RAN_CALL - Second Run

  Seed1         Seed2       n      x         y

523580480     911504187     1   0.24381   0.42445
192146343    1784774646     2   0.08948   0.83110
822902235     166306989     3   0.38319   0.07744
```

Functions That Generate Uniform Random Numbers

Function: RANUNI

Purpose: To generate a random number where each value is equally likely. The function generates uniform random numbers between 0 to 1 (note that the values 0 and 1 will never occur). By appropriate scaling, you can produce random numbers in any range you choose.

Note: UNIFORM is another name for this function.

Syntax: RANUNI (*seed*)

seed is a number less than 2,147,483,647 ($2^{31} - 1$). Every random number function needs to start with a seed, which is used to generate the first random number, and a new seed for the next time the function is executed. If the seed is less than or equal to 0, you will obtain a different series of random numbers each time you run your program (it uses the computer's clock to generate the seed). If you supply a positive seed, every time the program runs, it will produce the same series of random numbers.

Examples

Function	Returns
RANUNI(0)	Random number in the range 0 to 1 (clock-generated seed)
RANUNI(1357)	Random number in the range 0 to 1 (user-specified seed)

Program 9.3: Selecting an approximate *n*% random sample

```
***Primary function: RANUNI;

***This first DATA step generates a test data set;
data big;
   do Subj = 1 to 1000;
      output;
   .end;
run;

***This DATA step demonstrates how to select a random subset;
data random1;
   set big (where=(ranuni(456) le .10));
run;
```

Explanation

First, it is important to point out that the work of most of the programs in this chapter that select random subsets or assign subjects to groups can be accomplished with PROC SURVEYSELECT (part of SAS/STAT software). However, the programs are useful in demonstrating how to use SAS random functions.

Each time the second DATA step iterates, the RANUNI function generates a random number between 0 and 1. Since a non-zero seed is supplied, this program will generate the same series of random numbers (and therefore, the same subset) each time it is run. Since all random numbers between 0 and 1 are equally likely, approximately 10% of the random numbers will be less than or equal to .10, thus selecting an approximate 10% sample.

The next program demonstrates how to select a random sample that contains **exactly** 10% of the observations from the larger data set. Note that this program requires a sort and is not appropriate for use with large data sets. If you have a large data set and want a random subset, either use the program above to obtain an approximate subset or use PROC SURVEYSELECT.

Program 9.4: Selecting a random sample with exactly *n* observations

```
***Primary function: RANUNI;

data random2;
   set big;
   Shuffle = ranuni(0);
run;

proc sort data=random2;
   by Shuffle;
run;

data exact_n;
   set random2(drop=Shuffle obs=100);
run;
```

Explanation

The first DATA step makes a copy of the original data set (BIG) and adds a uniform random number to each observation. Sorting the data set by the random number, in effect, shuffles the order of the observations. Finally, by using the OBS= data set option, you can pick off the first 100 observations (or any number you choose) to create a subset with exactly *n* observations.

Program 9.5: Simulating random throws of two dice

```
***Primary function: RANUNI
***Other function: CEIL;

data dice;
   do i = 1 to 1000; /* generate 1000 throws */
      Die_1 = ceil(ranuni(0)*6);
      Die_2 = ceil(ranuni(0)*6);
      Throw = Die_1 + Die_2;
      output;
   end;
run;

title "Frequencies of dice throws";
proc freq data=dice;
   tables Die_1 Die_2 Throw / nocum nopercent;
run;
```

Explanation

The main point of this program is to demonstrate how to generate uniform random integers in a given range—in this case from 1 to 6. The RANUNI function returns random numbers **between** 0 and 1. So, when you multiply the RANUNI function by 6, the resulting values are between 0 and 6. The CEIL function returns the next largest integer (i.e., it rounds up). Therefore, CEIL(RANUNI(0)*6) gives you integers from 1 to 6. It's best not to use the ROUND function in programs like this, since you have to take extra precautions to ensure that each of the resulting values is equally likely.

The following listing from the FREQ procedure shows that you are producing integers in the proper range:

```
Frequencies of dice throws

The FREQ Procedure

Die_1    Frequency
         ─────────
   1        162
   2        178
   3        158
   4        176
   5        159
   6        167


Die_2    Frequency
         ─────────
   1        184
   2        179
   3        170
   4        151
   5        171
   6        145


Throw    Frequency
         ─────────
   2         28
   3         63
   4         89
```

(continued)

5	105
6	137
7	184
8	149
9	99
10	74
11	49
12	23

Program 9.6: Randomly assigning *n* subjects into two groups: Method 1—Approximate number of subjects in each group

```
***Primary function: RANUNI;

data assign1;
   do Subj = 1 to 12;
      if ranuni(123) lt .5 then Group = 'A';
    else Group = 'B';
    output;
   end;
run;

title "List of random assignments";
proc print data=assign1 noobs;
run;
```

Explanation

The DO loop generates 12 subject numbers (SUBJ). Since the RANUNI function generates uniform random numbers in the range 0 to 1, the function should return a value less than .5 about half the time. When this happens, GROUP is set to 'A'. If the number is greater than or equal to .5, GROUP gets set to 'B'. By the way, there is no need to worry that the program uses a less-than operator (LT) to assign a subject into group A and all others into B. The probability of returning a value exactly equal to .5 is so small that it can be ignored. When you use this method of random assignment, you are not guaranteed to assign equal numbers of subjects to groups A and B. If that is important to you, use the logic in the next program.

```
List of random assignments

Subj    Group

  1       B
  2       A
  3       A
  4       B
  5       A
  6       A
  7       B
  8       A
  9       A
 10       A
 11       B
 12       A
```

Program 9.7: Randomly assigning *n* subjects into two groups: Method 2— Equal number of subjects in each group

```
***Primary function: RANUNI;

proc format;
   value grpfmt 0 = 'A'  1 = 'B';
run;

data assign2;
   do Subj = 1 to 12;
      Group = ranuni(123);
      output;
   end;
run;

proc rank data=assign2 out=random groups=2;
   var Group;
run;

title "Random assignment of subjects, equal number in each group";
proc print data=random noobs;
   id Subj;
   var Group;
   format Group grpfmt.;
run;
```

Explanation

The DATA step creates a data set consisting of 12 observations. Each observation has a subject number and a variable called GROUP, which is a random number between 0 and 1. The heart of this program is PROC RANK. This procedure is commonly used to replace data values with their ranks. The smallest value is rank 1, the next smallest value rank 2, etc. When you add the GROUPS= option to PROC RANK, the procedure places all the observations into the number of groups you specify. In this program, where we have coded GROUPS=2, all values of GROUP below the median will be given the value of 0, and all values above the median will be given the value of 1. (Without the GROUPS= option, PROC RANK starts counting from 1; with the GROUPS= option, the groups begin at 0.)

So, the output data set created by PROC RANK now has a variable called GROUP with values of 0 and 1. The PRINT procedure applies the user-defined format GRPFMT to the GROUP variable so the listing consists of A's and B's. As long as there is an even number of subjects, this program can be used to randomly assign subjects to two groups with the same number of subjects in each group.

```
Random assignment of subjects, equal number in each group

Subj    Group

  1       B
  2       A
  3       A
  4       B
  5       A
  6       A
  7       B
  8       B
  9       A
 10       A
 11       B
 12       B
```

**Program 9.8: Randomly assigning _n_ subjects into two groups: Method 3—
Equal number of subjects in each group within blocks of four
subjects**

```
*** Primary function: RANUNI;

data assign3;
   do Block = 1 to 3;
      do i = 1 to 4;
         Subj + 1;
         Group = ranuni(123);
         output;
      end;
   end;
   drop i;
run;

proc rank data=assign3 out=random groups=2;
   by Block;
   var Group;
run;

title "Random assignment of subjects, blocks of four";
proc print data=random noobs;
   id Subj;
   var Group;
   format Group grpfmt.;
run;
```

Explanation

When you are assigning subjects to random groups, chance can deal you some strange
results. You may wind up with quite a few subjects in a row all in the same group. There are
times when you want to avoid this. This program assigns a random group to blocks of four
subjects. Thus, to randomly assign 12 subjects into two groups, you need to generate three
blocks of 4. With this method, you can be sure that you will not have too many subjects in a
row assigned to one group.

```
Random assignment of subjects, blocks of four

Subj    Group

  1       B
  2       A
  3       A
  4       B
  5       A
  6       A
  7       B
  8       B
  9       A
 10       A
 11       B
 12       B
```

Two Macros to Assign *n* Subjects into *k* Groups

For those who need to provide random assignments of subjects for experimental purposes, the following two macros are presented. The first macro performs a simple random assignment with an equal number of subjects in each group (providing that the total number of subjects is a multiple of the number of groups). The second macro performs the same function but goes one step further by randomizing subjects within blocks to attempt to avoid long runs of subjects assigned to the same group. Since these macros are included only as a convenience to those needing to perform this function, an explanation is not included.

Program 9.9: Macro to assign *n* subjects into *k* groups

```
***Primary function: RANUNI;

***Macro to assign n subjects into k groups;
*----------------------------------------------------------*
| Macro Name: ASSIGN                                       |
| Purpose: Random assignment of k treatments for n subjects |
| Arguments: Dsn=    : Data set to hold output             |
|            n=      : Number of subjects                  |
|            K=      : Number of groups                    |
|            Seed=   : Number for seed, seed is 0 if omitted |
|            Report= : NO if no report is desired          |
| Example: %assign (Dsn=mydata, n=100, k=4)                |
|          will assign 100 subjects into 4 equal groups    |
| Example: %assign (Dsn=mydata, n=100, k=4, seed=12345,    |
|                   Report=NO)                             |
|          will assign 100 subjects into 4 equal groups    |
|          Seed is 123 and no report is desired            |
*----------------------------------------------------------*;
%macro assign(Dsn=,    /* Data set to hold output        */
              n=,      /* Number of subjects             */
              k=,      /* Number of groups               */
              Seed=0,  /* Random number seed, 0 for random
                          value > 0 for fixed             */
              Report=YES /* Set to NO for no report       */);
   data _temp_;
      do Subj = 1 to &n;
         Group = ranuni(&Seed);
         output;
      end;
      run;
   proc rank data=_temp_ out=&Dsn groups=&k;
      var Group;
   run;
   ***increment group so group numbers start with 1;
   data &Dsn;
      set &Dsn;
      Group = Group + 1;
   run;
   %if %upcase(&Report) eq YES %then %do;
        title "&n of subjects randomly assigned to &k groups";
        proc report data=&Dsn nowd panels=99;
        columns Subj Group;
        define Subj / display width=4 right format=4.0 'Subj';
        define Group / display width=5 center format=2. 'Group';
     run;
   %end;
```

```
    proc datasets library=work;
       delete _temp_;
    run;
    quit;
 %mend assign;
```

To demonstrate this macro, I provide the following calling sequence to create a SAS data set called RANDOM, with 100 subjects assigned to four groups. In addition, the sequence uses a user-selected seed of 12345 and produces a report, listing the subject numbers and group assignments.

%*assign* (Dsn=mydata, n=**100**, k=**4**, seed=**12345**, Report=yes)

Here is a partial listing from this program:

```
100 of subjects randomly assigned to 4 groups

 Subj  Group     Subj  Group     Subj  Group     Subj  Group
   1     2         14    2         27    1         40    1
   2     3         15    2         28    1         41    4
   3     4         16    4         29    4         42    4
   4     1         17    2         30    1         43    3
   5     1         18    1         31    3         44    1
   6     3         19    1         32    1         45    2
   7     1         20    1         33    4         46    3
   8     3         21    2         34    4         47    3
   9     3         22    3         35    2         48    4
  10     4         23    2         36    1         49    1
  11     2         24    4         37    2         50    4
  12     1         25    3         38    1         51    1
  13     1         26    2         39    4         52    4
              etc.
```

Program 9.10: Macro to assign *n* subjects into *k* groups, with *b* subjects per block

```
***Primary function: RANUNI;

%macro assign_balance
    (Dsn=,         /* Data set to hold the output   */
     n=,           /* Number of subjects           */
     k=,           /* Number of groups             */
     b=,           /* Number of subjects per block */
     Seed=0,       /* Seed value, 0 is default     */
     report=yes /* Set to NO if no report wanter */);

/* Example:
    %assign_balance(Dsn=mydata, n=100, k=2, b=10)
    assign 100 subjects into two groups blocked
    every 10 subjects
*/
    data _temp_;
    %if %eval(%sysevalf(&b/&k, floor) - %sysevalf(&b/&k, ceil))
        ne 0 %then %do;
        file print;
        put "The number of subjects per group (&b) is not"/
            "evenly divisible by the number of groups (&k)";
        stop;
    %end;
        %let n_blocks = %sysevalf(&n/&b,floor);
        do Block = 1 to &n_blocks;
            do j = 1 to &b;
                Subj + 1;
                Group = ranuni(&seed);
                output;
            end;
        end;
    run;
    proc rank data=_temp_ out=&Dsn Groups=&k;
        by Block;
        var Group;
    run;
    ***increment group so group numbers start with 1;
    data &Dsn;
        set &Dsn(drop = Block);
        Group = Group + 1;
    run;
    %if %upcase(&Report) eq YES %then %do;
    title1 "&n subjects randomly assigned to &k groups";
    title2 "Equal number of subjects in each group of &b subjects";
```

```
proc report data=&dsn nowd panels=99;
   columns Subj Group;
   define Subj / display width=4 right format=4.0 'Subj';
   define Group / display width=5 center format=2. 'Group';
run;
%end;
proc datasets library=work;
   delete _temp_;
run;
quit;
%mend assign_balance;
```

Function: CALL RANUNI

As discussed in the introduction to this chapter, all of the random number functions have a corresponding CALL routine.

Purpose: To generate one or more random numbers where each value is equally likely. The function generates uniform random numbers in the range from 0 to 1. By appropriate scaling, you can produce random numbers in any range you choose.

Syntax: CALL RANUNI(*seed, value*)

seed is an integer less than 2,147,483,647 ($2^{31} - 1$). If the seed is less than or equal to 0, you obtain a different series of random numbers each time you run your program (it uses the computer's clock to generate the seed). If you supply a positive seed, every time the program runs, it will produce the same series of random numbers. You need to initialize the seed before you call RANUNI for the first time—it cannot be a constant.

value is the name of a variable that will hold the uniform random number.

Examples

For these examples, SEED1 = 0 and SEED2 = 1234567.

Function	Returns
CALL RANUNI(SEED1, X)	A value for X in the range 0 to 1 (clock-generated seed)
CALL RANUNI(SEED2, Y)	A value for Y in the range 0 to 1 (user-specified seed)

See Program 9.2 for an example of the RANUNI CALL routine.

Functions That Generate Normally Distributed Numbers

Function: RANNOR

Purpose: To generate random numbers that are normally distributed. This function generates random numbers that come from a distribution with a mean of 0 and a standard deviation of 1. You can easily scale this to generate any normally distributed values. For example, to generate normally distributed numbers with a mean of 100 and a standard deviation of 10, you multiply the result of the RANNOR function by 10 and add 100.

Syntax: RANNOR (*seed*)

seed is either a 0, in which case you will obtain a different series of normally distributed random numbers each time you run your program (it uses the computer's clock to generate the seed), or a number of your choosing, in which case every time the program runs, it produces the same series of random numbers.

Examples

Function	Returns
RANNOR(0)	A random number from a normal distribution (mean = 0, standard deviation = 1) with a machine-generated seed
RANNOR(1234)	A random number from a normal distribution (mean = 0, standard deviation = 1) with user-defined seed
10*RANNOR(0) + 100	A random number from a normal distribution (mean = 100, standard deviation = 10) with a machine-generated seed

The program that follows uses the RANNOR function to demonstrate how to run a Monte Carlo simulation using SAS. Two groups are defined with their own mean, standard deviation, and sample size. The program then generates 1,000 sets of two samples, runs 1,000 *t*-tests on the 1,000 samples, and uses the Output Delivery System (ODS) to capture the *p*-values from each of these 1,000 samples. From this, the power of the *t*-tests can be determined. This same technique can be used with other distributions or other statistical tests.

Program 9.11: Demonstrating a Monte Carlo simulation to determine the power of a *t*-test

```
***Primary function: RANNOR;

/***************************************************************
    Group 1:mean = 100 n = 8 standard deviation = 10
    Group 2:mean = 115 n = 6 standard deviation = 12
    Scores are normally distributed in each group.
    This program generates 1000 samples;
****************************************************************/
data generate;
   do Block = 1 to 1000;
      do Group = 'A', 'B';
         if group = 'A' then do Subj = 1 to 6;
            x = rannor(123)*10 + 100;
            output;
         end;
         else if group = 'B' then do subj = 1 to 8;
            x = rannor(123)*12 + 115;
            output;
         end;
      end;
   end;
run;

***Test the distributions;
title "Mean and Standard Deviation of the 1000 Samples";
proc means data=generate n mean std;
   class Group;
   var x;
run;

***run 1000 t-tests and capture the p-values;
ods listing close;
ods output ttests=work.p_values;
```

```
proc ttest data=generate;
   by Block;
   class Group;
   var x;
run;
ods output close;
ods listing;

***Examine the results;
data power_t;
   set p_values;
   if Probt le .05 then result = 'Power';
   else Result = 'Beta';
run;
title "Power of a t-test for assumptions of equal or unequal variance";
proc tabulate data=power_t;
   class Variances Result;
   tables Variances , (Result all)*pctn<Result all>=' ';
run;
title "Power of a t-test with equal variances";
proc chart data=p_values;
   where Variances = 'Equal';
   vbar Probt / midpoints = 0 to .7 by .05;
run;
quit;
```

Explanation

The first DATA step generates the 1,000 samples (the data set BLOCKS). Remember that a DO loop can use character values as well as numerical ones. Here, GROUP is first set to 'A'. Then six values of X are generated where X has a mean of 100 and a standard deviation of 10. The next time through this inner loop, GROUP gets set to 'B', and eight values of X are generated where X has a mean of 115 and a standard deviation of 12.

Just to be sure that this DATA step is working properly, PROC MEANS is run to determine the mean and standard deviation of the 1,000 samples of GROUP 'A' and GROUP 'B'. This output is shown next:

```
Mean and Standard Deviation of the 1000 Samples

The MEANS Procedure
                    Analysis Variable : x
Group    N Obs      N          Mean         Std Dev

A        6000      6000      99.6789555      10.0285852

B        8000      8000     115.0191877      11.9847446
```

Notice that these values are close to the specified values.

Next, two ODS statements are submitted. The first closes the listing file so that the output window doesn't fill up. The next ODS statement creates a SAS data set called P_VALUES, containing part of the *t*-test output containing the *p*-values. To determine what goes into this data set, you need to know the names of the various output objects from each procedure. To do that, you can run the desired procedure once, preceded by the ODS TRACE statement with the ON argument to obtain a list of the output objects from that procedure. With PROC TTEST, the output object you want is called TTESTS. You will need to run a PROC PRINT on this data set to see exactly what it looks like.

Having done that, you will find that the *p*-values from each of the 1,000 *t*-tests were stored in a variable called PROBT. Therefore, for each sample where the *p*-value was less than or equal to .05, the variable RESULT was assigned a value of Power. Otherwise it was set to Beta. Running a simple PROC FREQ step would have sufficed to determine the percentage of samples where RESULT was equal to Power (the power of the test) or Beta (the beta error). In this case, the fancier listing from PROC TABULATE was used to make a prettier listing. It is shown here:

```
Power of a t-test for assumptions of equal or unequal variance
```

	result		
	Beta	Power	All
Variances			
Equal	35.40	64.60	100.00
Unequal	34.60	65.40	100.00

Finally, a power curve is displayed by running PROC CHART.

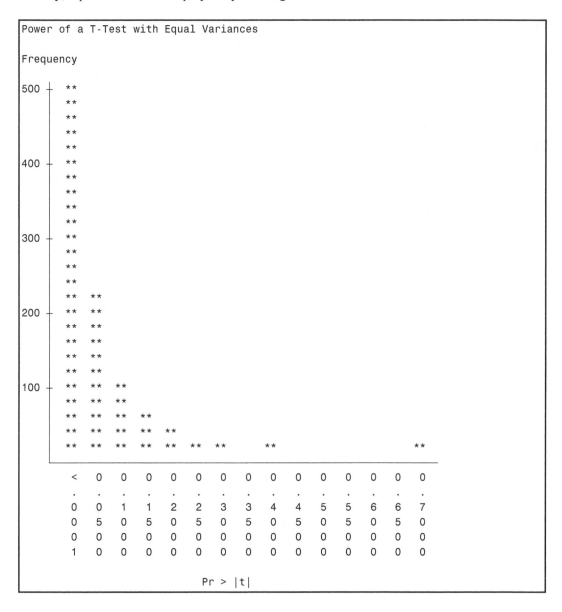

Function: CALL RANNOR

Purpose: To generate random numbers that are normally distributed. This CALL routine generates random numbers that come from a distribution with a mean of 0 and a standard deviation of 1. You can easily scale this to generate any normally distributed values. For example, to generate normally distributed numbers with a mean of 100 and a standard deviation of 10, you would multiply the result of the RANNOR function by 10 and add 100.

Syntax: CALL RANNOR(*seed, value*)

seed is either a 0, in which case you obtain a different series of normally distributed random numbers each time you run your program (it uses the computer's clock to generate the *seed*), or a number of your choosing, in which case *seed* produces the same series of random numbers every time the program runs.

value is the name of a variable that holds the normally distributed random number.

Examples

For these examples, SEED1 = 0 and SEED2 = 1234567.

Function	Returns
CALL RANNOR(SEED1, X)	A value for X from a normal distribution with a mean of 0 and a standard deviation of 1 (clock-generated seed)
CALL RANNOR(SEED2, Y)	A value for Y from a normally distributed distribution with a mean of 0 and a standard deviation of 1 (user-specified seed)

Program 9.12: Program to generate random values of heart rates that are normally distributed

```
***Primary function: RANNOR;
***Other function: ROUND;

data generate;
   Seed = 0;
   do Subj = 1 to 100;
      call rannor(Seed, HR);
      HR = round(15*HR + 70);
      output;
   end;
run;

options ps=16;
title "Listing of Data Set GENERATE";
proc report data=generate panels=99 nowd;
   columns Subj HR;
   define Subj / display width=4;
   define HR / display width=4 format=4.0;
run;
```

Explanation

The values for heart rate (HR) are generated from a normal distribution with a mean of 70 and a standard deviation of 15. Since *seed* is set to 0, you will obtain a different series of heart rates every time you run the program. PROC REPORT was used with the PANELS option so that the output could be shown in multi-column format. The listing is shown next:

```
                        Listing of Data Set GENERATE

      SUBJ    HR     SUBJ    HR     SUBJ    HR     SUBJ    HR
       53     82      66     67      79     66      92     70
       54     78      67     69      80     59      93     85
       55     61      68     79      81     72      94     86
       56     56      69     79      82     58      95    112
       57     79      70     73      83     80      96     75
       58     60      71     56      84     66      97     69
       59     99      72     85      85     65      98     88
       60     47      73     88      86     83      99     79
       61    101      74     75      87     87     100     66
       62     95      75     77      88     64
       63     56      76     45      89     92
       64     88      77     79      90     42
       65     60      78     66      91     44
```

A Function That Generates Random Numbers from a Wide Variety of Probability Distributions

Function: RAND

Purpose: To generate random numbers from any of dozens of probability distributions, including Uniform, Normal, and Binomial (see the following complete list).

This function can perform all the actions of RANUNI and RANNOR as well as the other probability distributions. You may want to use this function instead of either of the two alternatives, since developers at SAS Institute feel that it does a better job of generating random values than its predecessors. According to the *SAS 9.2 Language Reference: Dictionary, Second Edition* (2009):

The RAND function uses the Mersenne-Twister random number generator (RNG) that was developed by Matsumoto and Nishimura (1998). The random number generator has a very long period (219937 – 1) and very good statistical properties.

Important Note: If you want a reproducible series of random numbers using the RAND function, you must seed it by a call to STREAMINIT (with a positive integer argument) prior to its use. For example:

```
call streaminit(132435);
```

Place this call only once in the DATA step before the first use of the RAND function in the step.

CALL STREAMINIT(0) (or negative values) will generate random values where the CPU clock is used to generate the first seed value (thus generating a different series of random numbers each time the program is run.) This is the default action if you do not call STREAMINIT.

Syntax: RAND(*distribution-name* <,*parameters*>)

distribution-name is the name of a probability distribution chosen from the following list:

Distribution	1st Argument in Function
Bernoulli	BERNOULLI
Beta	BETA
Binomial	BINOMIAL
Cauchy	CAUCHY
Chi-Square	CHISQUARE
Erlang	ERLANG
Exponential	EXPONENTIAL
F	F
Gamma	GAMMA
Geometric	GEOMETRIC
Hypergeometric	HYPERGEOMETRIC
Lognormal	LOGNORMAL
Negative binomial	NEGBINOMIAL
Normal	NORMAL|GAUSSIAN
Poisson	POISSON
T	T
Tabled	TABLE
Triangular	TRIANGLE
Uniform	UNIFORM
Weibull	WEIBULL

Some of these distributions allow you to supply parameters. We only present the parameters for some of the more popular distributions. For a complete list, please see the Product Documentation in the Knowledge Base, available at http://support.sas.com/documentation.

Distribution: Uniform
Optional Parameters: none
Example: RAND('Uniform') generates uniform random numbers between 0 and 1.

Distribution: Normal
Optional Parameters: m (mean), s (standard deviation)
Example: RAND('Normal',100,5) generates random numbers that are normally distributed with a mean of 100 and a standard deviation of 5.

Distribution: Bernoulli
Optional Parameter: p (probability of success ($0 \leq p \leq 1$)
Example: RAND('Bernoulli',.4) generates 0's and 1's with the probability of a 1 equal to .4.

Distribution: Binomial
Optional Parameters: p (probability of success), n (the number of Bernoulli trials)
Example: RAND('Binomial',.2, 5) generates random numbers that represent the number of successes in a sample of size 'n' where the probability of success is equal to .2.

Distribution: Poisson
Optional Parameters: m (mean)
Example: RAND('Poisson',5) generates random integers that represent the number of successes when the mean of the Poisson distribution is 5.

Distribution: T
Optional Parameters: df (degrees of freedom)
Example: RAND('T',10) generates random numbers from a t-distribution with 10 degrees of freedom.

Program 9.13: Generating a series of normally distributed values using the RAND function

```
***Primary function: RAND;
data normal;
   call streaminit(13245);
   do i = 1 to 10;
      x = rand('Normal',100,5);
      output;
   end;
   drop i;
run;

title "Listing of data set NORMAL";
proc print data=normal noobs;
run;
```

This program generates 10 numbers that are normally distributed with a mean of 100 and a standard deviation of 5. It you run this same DATA step a second time, you will obtain the same series of random values.

```
Listing of data set NORMAL

   x

 85.149
101.819
 98.728
 98.176
 93.499
 99.093
101.969
 97.594
101.149
104.345
```

References

Matsumoto, M., and T. Nishimura. 1998. "Mersenne Twister: A 623-Dimensionally Equidistributed Uniform Pseudo-Random Number Generator." *ACM Transactions on Modeling and Computer Simulation* 8: 3–30.

SAS Institute Inc. 2009. *SAS 9.2 Language Reference: Dictionary, Second Edition*. Cary, NC: SAS Institute Inc. Available at http://support.sas.com/documentation.

C h a p t e r   1 0

Special Functions

Introduction

The title of this chapter may cause you to think that these functions are somehow esoteric or unimportant. In fact, these functions are some of the most commonly used functions in the SAS arsenal—they just don't fit easily into any of the other categories of SAS functions.

Functions That Obtain Values from Previous Observations

These two functions are grouped together because of their similar applications. The LAG function returns the value of its argument the last time the function executed. If you execute the LAG function in every iteration of the DATA step, you can interpret the result of the LAG function as the value of its argument in the previous observation. There is actually a whole class of LAG and DIF functions—LAG1, LAG2, etc.—which return the value from the previous iteration of the DATA step, the iteration before that, etc. Each of the family of DIF functions operates in a similar manner to the family of LAG functions except that the difference between the current value and a previous value is returned. For example, DIF(X) returns the current value of X minus the value of X the last time the DIF function executed. Thus, DIF is very useful in determining the difference between a value in the current observation and a value in a previous observation.

Functions: LAG and LAG*n*

Purpose: To obtain the value of a variable from a previous observation or *n* observations previous to the current observation.

Caution: This application of the LAG and LAG*n* functions requires that you execute the function in **every** iteration of the DATA step. If you conditionally execute a LAG*n* function, the results you get may be unexpected and unwanted.

Syntax: LAG(*value*)
LAG*n*(*value*)

value is any numeric variable.

n is the number of lagged values. LAG and LAG1 are equivalent.

Examples

For these examples, the value of X in the current observation is 9, the value of X in the previous observation is 7, and the value of X in the observation previous to that is 5.

Function	Returns
LAG(X)	7
LAG2(X)	5
LAG3(X)	missing

Program 10.1: Using the LAG*n* functions to compute a moving average

```
***Primary functions: LAG and LAG2
***Other function: MEAN;

***Program to compute a moving average, based on three observations;
data moving;
   input x @@;
   x1 = lag(x);
   x2 = lag2(x);
   Moving = mean(x, x1, x2);
   if _n_ ge 3 then output;
datalines;
1 3 9 5 7 10
;
title "Listing of Data Set MOVING";
proc print data=moving noobs;
run;
```

Explanation

In the first iteration of the DATA step, X has a value of 1; X1 and X2 are both missing. During the second iteration of the DATA step, X has a value of 3, X1 is equal to 1, and X2 is missing. Finally, in the third iteration of the DATA step, X is equal to 9, X1 is equal to 3, and X2 is equal to 1. Since _N_ is now equal to 3, the OUTPUT statement is executed, and the first observation in the data set MOVING is written out. See the following listing:

```
Listing of Data Set MOVING

 x    x1    x2     Moving

 9     3     1     4.33333
 5     9     3     5.66667
 7     5     9     7.00000
10     7     5     7.33333
```

Functions: DIF and DIF*n*

Purpose:　To compute the difference between a value in the current observation and a value from one or more previous observations. Note that DIF(X) is equivalent to X - LAG(X).

Syntax:　DIF(*value*)
　　　　　DIFn(*value*)

　　　　value is any numeric variable.

　　　　n is the number of lagged values. LAG and LAG1 are equivalent.

Examples

For these examples, the value of X in the current observation is 9, the value of X in the previous observation is 7, and the value of X in the observation previous to that is 5.

Function	Returns
DIF(X)	2
DIF2(X)	4
DIF3(X)	.(missing)

Program 10.2: Computing changes in blood pressure from one visit to another

```
***Primary function: DIF;
***Create a test data set of patient visits;

data visits;
   input ID Visit_date : mmddyy10. SBP DBP @@;
   format Visit_date date9.;
   label SBP = 'systolic blood pressure'
         DBP = 'diastolic blood pressure';
datalines;
1 02/01/2003 180 110   1 03/02/2003 178 100   1 04/01/2003 170 90
2 03/03/2003 170 100   2 04/01/2003 172 100
3 04/01/2003 130 80    3 06/01/2003 128 82    3 08/01/2003 128 78
;
proc sort data=visits;
   by ID Visit_date;
run;
```

```
***Program to compute changes between visits;
data change;
   set visits;
   by ID;

   ***Delete any subject with only one visit;
   if first.ID and last.ID then delete;

   Diff_SBP = dif(SBP);
   Diff_DBP = dif(DBP);
   if not first.ID then output;
run;

title "Listing of Data Set CHANGE";
proc print data=change noobs;
run;
```

Explanation

Notice that the DIF function is executed during every iteration of the DATA step. For the first visit for subject one (the first observation in VISITS), the value of the DIF function will be missing (the current value minus a missing value). For the second visit, the DIF function will take the current value of SBP (178) minus the value from the previous visit (180) and return a value of –2. Now, what happens when the DATA step reaches the first visit for the second subject? It is actually computing the difference between the current value of SBP (170) and the value of SBP from the previous subject! This is OK. You need to execute the DIF function for every iteration of the DATA step. However, because of the IF NOT FIRST.ID statement (a subsetting IF statement), the DATA step does not output an observation when it is processing the first visit for every patient. You need to delete any subject with only one visit or you will be computing the difference between his or her values and the last value from the previous subject. Notice that the values for the first visit for each patient are missing from the following listing. (To learn more about FIRST. and LAST. variables, check out *Longitudinal Data and SAS: A Programmer's Guide*, by this author.)

```
Listing of Data Set CHANGE

         Visit_
ID         date    SBP    DBP    Diff_SBP    Diff_DBP

 1      02MAR2003   178    100      -2          -10
 1      01APR2003   170     90      -8          -10
 2      01APR2003   172    100       2            0
 3      01JUN2003   128     82      -2            2
 3      01AUG2003   128     78       0           -4
```

**Program 10.3: Computing the difference in blood pressure between the
first and last visit for each patient**

```
***Primary function: DIF;
***This example uses the data set VISITS from the example above;

data first_last;
   set visits;
   by ID;
   ***Note: The DIF function is being executed conditionally.
      Be VERY careful if you do this;
   ***Delete any subject with only one visit;
   if first.ID and last.ID then delete;

   if first.ID or last.ID then do;
      Diff_SBP = dif(SBP);
      Diff_DBP = dif(DBP);
   end;
   if last.ID then output;
run;

title "Listing of Data Set FIRST_LAST";
proc print data=first_last noobs;
run;
```

Explanation

This example clearly demonstrates what happens when you conditionally execute the DIF
(or LAG) function. Notice that the DIF function is executed only for the first or last visit for
each subject. As before, the value obtained during the first visit is either the current value
minus a missing value (first subject) or the current value minus the last value for the
previous subject. This is necessary to "prime the pump" and ensure that the next time the
DIF function is executed, it will be computing the difference between the first and last visit.

Notice also that an observation is output only when the DATA step is processing the last visit for each subject. This is a very unusual application of the DIF function and you should always use extreme care if the DIF or LAG functions are executed conditionally. As before, you need to omit the values from any subject who has only one visit. The following listing shows one observation for each subject, with the current value of SBP and DBP and the difference between the first and last visit.

```
Listing of Data Set FIRST_LAST

        Visit_
ID        date     SBP    DBP    Diff_SBP    Diff_DBP

 1    01APR2003    170     90       -10         -20
 2    01APR2003    172    100         2           0
 3    01AUG2003    128     78        -2          -2
```

Functions That Perform Character-to-Numeric or Numeric-to-Character Conversion

Function: INPUT

The first two functions in this group (INPUT and PUT) are being grouped together since they have something in common: they are commonly used to perform character-to-numeric (INPUT) or numeric-to-character (PUT) conversion. These two functions may be confusing at first, but they are extremely useful once you get the hang of them.

Purpose: To perform character-to-numeric conversion. Also useful in converting character values such as dates into true SAS numeric date values.

Syntax: INPUT(*value, informat*)

value is a character variable or character expression.

informat is a SAS or user-defined informat.

Examples

For these examples, N_CHAR = '123' and CHAR_DATE = '10/21/1980'.

Function	Returns
INPUT(N_CHAR,3.)	123 (numeric)
INPUT(N_CHAR,9.)	123 (numeric)
INPUT(CHAR_DATE,MMDDYY10.)	7599 (equals 10/21/1980)

Program 10.4: Using the INPUT function to perform character-to-numeric conversion

```
***Primary function: INPUT;
***Create test data set;

data char;
    input Num $ Date1 : $10. Date2 : $9. Money : $12.;
datalines;
123 10/21/1980 21OCT1980 $123,000.45
XYZ 11/11/2003 01JAN1960 $123
;
data convert;
    set char(rename=(Num=C_num));
    Num = input(C_num,9.);
    SASdate1 = input(Date1,mmddyy10.);
    SASdate2 = input(Date2,date9.);
    Dollar = input(Money,comma12.);
    format SASdate1 SASdate2 mmddyy10.;
run;

title "Listing of Data Set CONVERT";
proc print data=convert noobs;
run;
```

Explanation

All the variables in the CHAR data set are character variables. Although you can force SAS to do a character-to-numeric conversion by multiplying the character variable by 1 or adding 0, that causes SAS to write messages to the log, and is considered by most SAS programmers to be sloppy programming. In this example, the user wanted to use the same variable name (NUM) for the numeric variable that was in the CHAR data set as a character variable. The usual trick is to use the RENAME= data set option to rename the character variable, which allows you to keep the same name for the numeric equivalent. A SAS variable cannot be both a character and a numeric value at the same time.

The two dates were read as character values (this happens frequently when you are importing dates from various databases or spreadsheets). The INPUT function allows you to "reread" these values using the correct SAS date informats. Finally, the value with dollar signs and commas is converted to a numeric value by using the comma informat. Although not readily apparent, the dates in the following listing are true SAS dates that have been formatted using the MMDDYY10. format.

```
Listing of Data Set CONVERT

C_num      Date1          Date2        Money

 123       10/21/1980     21OCT1980    $123,000.45
 XYZ       11/11/2003     01JAN1960    $123

Num        SASdate1       SASdate2       Dollar

123        10/21/1980     10/21/1980     123000.45
  .        11/11/2003     01/01/1960        123.00
```

If you examine the SAS log (shown next), you will see a note there that the character value of XYZ could not be read with a numeric informat.

```
NOTE: Invalid argument to function INPUT at line 203 column 10.
C_NUM=XYZ DATE1=11/11/2003 DATE2=01JAN1960 MONEY=$123 NUM=.
SASDATE1=11/11/2003 SASDATE2=01/01/1960 DOLLAR=123 _ERROR_=1 _N_=2
NOTE: Mathematical operations could not be performed at the following
      places. The results of the operations have been set to missing
      values.
      Each place is given by: (Number of times) at (Line):(Column).
      1 at 203:10
```

To avoid error messages in the SAS log, you can use a ?? modifier with the INPUT function, much in the same way as you can with the INPUT statement. So, if you rewrote the line in question like this, there would be no errors listed in the SAS log.

```
INPUT (C_NUM,?? 9.);
```

Function: **INPUTC**

The next two functions in this group (INPUTC and INPUTN) are similar to the INPUT function with one important difference: They can assign the informat at run time. The INPUTC function is used for character informats and the INPUTN function is used with numeric formats. These functions are useful when you want to select an informat from information obtained from the data itself.

Purpose: Similar to the INPUT function except that the character informat can be assigned at run time.

Syntax: INPUTC(*value, char-informat*)

value is a character variable or character expression.

char-informat is a SAS or user-defined informat.

Examples

For these examples, INFOR1 = "$GENDER.", INFOR2 = "$YESNO.", VALUE1 = "1", and VALUE2 = "2" ($GENDER AND $YESNO are user defined informats).

Function	Returns
INPUTC(VALUE1, INFOR1)	"Male"
INPUTC(VALUE2, INFOR2)	"Yes"

Program 10.5: **Using the INPUTC function to specify an informat at run time**

```
***Primary function: INPUTC
***Other functions: PUT;

proc format;
   invalue $codea 'A' = 'Chair'
                  'B' = 'Desk'
                  'C' = 'Table';
   invalue $codeb 'A' = 'Office Chair'
                  'B' = 'Big Desk'
                  'C' = 'Coffee Table';
   value code 1 = '$codea.'
              2 = '$codeb.';
run;
```

```
data items;
   input Year Letter : $1. @@;
   length Item $ 12;
   Item = inputc(Letter, put(Year, code.));
datalines;
1 A  1 B  2 A  1 C  2 C  2 B
;
title "Listing of Data Set ITEMS";
proc print data=items noobs;
run;
```

Explanation

In this example, there are two sets of codes for office furniture, described by the two informats, $CODEA and $CODEB. The variable YEAR, read in as a data value, tells the program which informat to use. If YEAR is equal to 1, the result of PUT(YEAR, CODE.) is the character value "$CODEA." If YEAR is 2, the PUT function returns "$CODEB." This value is then used in the INPUTC function. Reading an "A" using the $CODEA informat results in the value "Chair." Reading an "A" using the $CODEB informat results in the value "Office Chair," etc. Here is a listing of data set ITEMS:

```
Listing of Data Set ITEMS

Year    Letter    Item

 1        A       Chair
 1        B       Desk
 2        A       Office Chair
 1        C       Table
 2        C       Coffee Table
 2        B       Big Desk
```

Function:　INPUTN

Purpose:　Similar to the INPUT function except that the numeric informat can be assigned at run time.

Syntax:　INPUTN(*value, num-informat*)

　　　　　　value is a character variable or character expression.

　　　　　　num-informat is a SAS or user-defined informat.

Examples

For these examples, DATE1 = "10/21/1980", DATE2 = "21OCT1980", INFOR1 = "MMDDYY10..", and INFOR2 = "DATE9.".

Function	Returns
INPUTN(DATE1, INFOR1)	7599
INPUTN(DATE2, INFOR2)	7599

Program 10.6:　Using the INPUTN function to read dates in mixed formats

```
***Primary function: INPUTN
***Other function: PUT;

proc format;
   value which 1 = 'mmddyy10.'
               2 = 'date9.';
run;

data mixed_dates;
   input Which_one Dummy : $10.;
   Date = inputn(Dummy, put(Which_one, which.));
   format Date weekdate.;
datalines;
1 10/21/1980
2 21OCT1980
1 01/01/1960
2 03NOV2003
;
title "Listing of Data Set MIXED_DATES";
proc print data=mixed_dates noobs;
run;
```

Explanation

Although there are simpler ways to solve this problem of mixed-date formats (using the ANYDTDTE informat, for example), this solution helps demonstrate the INPUTN function. The informat you use as the second argument of the INPUTN function is obtained by the result of the PUT function and the WHICH format. When the variable WHICH_ONE is equal to 1, the statement `PUT(WHICH_ONE, WHICH.)` returns the character value "MMDDYY10." When WHICH_ONE is equal to 2, the PUT function returns the value "DATE9." The INPUTN function uses these values at run time to determine which informat to use to read the date properly. The data set MIXED_DATES is shown next:

```
Listing of Data Set MIXED_DATES

Which_
  one      Dummy                   Date

   1     10/21/1980     Tuesday, October 21, 1980
   2     21OCT1980      Tuesday, October 21, 1980
   1     01/01/1960       Friday, January 1, 1960
   2     03NOV2003      Monday, November 3, 2003
```

Function: **PUT**

Purpose: To perform numeric-to-character conversion or to create a character variable from a user-defined format. The result of a PUT function is always a character value.

Note: The VVALUE function (that returns the formatted value of a variable) may be an alternative to the PUT function. Please refer to Chapter 15, "Variable Information Functions," for a description of this function.)

Syntax: `PUT(value, format)`

value is a character or numeric SAS variable or expression.

format is a SAS or user-defined format.

Examples

For these examples, `X = 3` (numeric), `DATE = "21OCT1980"`.

Function	Returns
PUT(X,3.)	"3" (character)
PUT(7599,MMDDYY10.)	"10/21/1980"
PUT(INPUT(DATE, DATE9.), MMDDYY10.)	"21OCT1980"
PUT(X, WEEKDATE3.)	"Mon"

Program 10.7: Performing a table look-up using a format and the PUT function

```
***Primary functions: PUT, INPUT;

proc format;
   value item 1 = 'APPLE'
               2 = 'PEAR'
               3 = 'GRAPE'
               other = 'UNKNOWN';
   value $cost 'A' - 'C' = '44.45'
               'D'       = '125.'
               other     = ' ';
run;

data table;
   input Item_no Code $ @@;
   Item_name = put(Item_no, item.);
   Amount = input(put(Code, $cost.),9.);
datalines;
1 B    2 D    3 X    4 C
;
title "Listing of Data Set TABLE";
proc print data=table noobs;
run;
```

Explanation

In order to associate an item name with the item number, the PUT function takes the formatted value of the item number and assigns this value to the character variable ITEM_NAME. The result of using the PUT function with the variable CODE are the character values defined by the $COST format. Since the result of a PUT function is always a character value, in order to obtain the amount as a numeric value, the INPUT function performs a character-to-numeric conversion. In the following listing, the variable AMOUNT is a numeric value.

```
Listing of Data Set TABLE

                    Item_
Item_no    Code     name       Amount

   1        B       APPLE       44.45
   2        D       PEAR       125.00
   3        X       GRAPE         .
   4        C       UNKNOWN     44.45
```

Function: PUTC

Just as with the INPUTC and INPUTN functions, the last two functions in this group (PUTC and PUTN) allow you to specify a character or numeric format at run time, respectively.

Purpose: Similar to the PUT function except that the character format can be assigned at run time.

Syntax: PUTC(*value, char-format*)

value is a character variable or character expression.

char-format is a SAS or user-defined character format.

Examples

For these examples, FOR1 = "$GENDER.", FOR2 = "$YESNO.", VALUE1 = "1", and VALUE2 = "2" ($GENDER and $YESNO are user-defined character formats).

Function	Returns
PUTC(VALUE1, FOR1)	"Male"
PUTC(VALUE2, FOR2)	"Yes"

Program 10.8: Using the PUTC function to assign a value to a character variable at run time

```
***Primary function: PUTC
***Other function: PUT;

proc format;
   value $tool '1' = 'Hammer'
               '2' = 'Pliers'
               '3' = 'Saw';
   value $supply '1' = 'Paper'
                 '2' = 'Pens'
                 '3' = 'Paperclips';
   value type 1 = '$TOOL.'
              2 = '$SUPPLY.';
run;

data tools_supplies;
   input Type Value $;
   length Name $ 10;
   Format = put(Type, type.);
   Name = putc(Value, Format);
datalines;
1 1
2 1
1 2
2 3
;
title "Listing of Data Set TOOLS_SUPPLIES";
proc print data=tools_supplies noobs;
run;
```

Explanation

In this program, the PUT function, along with a format, assigns the proper value to the variable FORMAT. This, in turn, is used in the PUTC function to supply the proper character format (either $TOOLS or $SUPPLY) to use in the PUTC function. A listing of data set TOOLS_SUPPLIES follows:

```
Listing of Data Set TOOLS_SUPPLIES

Type    Value    Name          Format

  1       1      Hammer        $TOOL.
  2       1      Paper         $SUPPLY.
  1       2      Pliers        $TOOL.
  2       3      Paperclips    $SUPPLY.
```

Function: PUTN

Purpose: Similar to the PUT function except that the numeric format can be assigned at run time.

Syntax: PUTN(value, numeric-format)

value is a numeric variable or expression.

numeric-format is a SAS or user-defined numeric format.

Examples

For these examples, FOR1 = "GENDER.", FOR2 = "YESNO.", VALUE1 = 1, and VALUE2 = 2 (GENDER and YESNO are user-defined numeric formats).

Function	Returns
PUTN(VALUE1, FOR1)	"Male"
PUTN(VALUE2, FOR2)	"Yes"

Program 10.9: Using the PUTN function to assign a value to a character variable at run time

```
***Primary function: PUTN;

proc format;
   value tool 1 = 'Hammer'
              2 = 'Pliers'
              3 = 'Saw';
   value supply 1 = 'Paper'
                2 = 'Pens'
                3 = 'Paperclips';
run;
data tools_supplies;
   input Type $ Value;
   Name = putn(Value, Type);
datalines;
TOOL. 1
SUPPLY. 1
TOOL. 2
SUPPLY. 2
;
title "Listing of Data Set TOOLS_SUPPLIES";
proc print data=tools_supplies noobs;
run;
```

Explanation

To keep this program very simple, the actual format (TOOL or SUPPLY) was read into the variable TYPE. This value was then used in the PUTN function to translate the numeric values into either a list of tools or a list of supplies. A listing of TOOLS_SUPPLIES is shown next:

```
Listing of Data Set TOOLS_SUPPLIES

 Type      Value    Name

TOOL.        1      Hammer
SUPPLY.      1      Paper
TOOL.        2      Pliers
SUPPLY.      2      Pens
```

Function That Sets Variable Values to a SAS Missing Value

CALL MISSING is an extremely useful CALL routine that sets the value of its arguments to a SAS missing value. It takes both numeric and character variables.

This CALL routine is most useful when you are writing a program where some of the variables are retained and you need to initialize their values to missing at various times in the DATA step.

Function: **CALL MISSING**

Purpose: To set all of its arguments (numeric or character) to a SAS missing value.

Syntax: `CALL MISSING(<of> Variable-1 <,…Variable-n>))`

`Variable` is a numeric or character variable.

If you use any form of a variable list such as x1–x3, _NUMERIC_, _CHARACTER_, or _ALL_, you need to precede the variable list with the keyword "of".

Examples

For these examples, x1=9, x2=8, x3=., c1='Ron', c2='Cody'.
An array statement: `array x[3];` was also defined.

Function	Returns
CALL MISSING(x1,x2,x3)	x1=., x2=., x3=.
CALL MISSING(of x1-x3)	x1=., x2=., x3=.
CALL MISSING(of x[*])	x1=., x2=., x3=.
CALL MISSING(of _numeric_)	x1=., x2=., x3=.
CALL MISSING(of _all_)	x1=., x2=., x3=., c1=' ', c2=' '

Program 10.10: Demonstrating the CALL MISSING routine

```
***Primary function: CALL MISSING;
title;
data _null_;
   input x1-x3 (Char1-Char3) (: $5.);
   file print;
   put "Values before the CALL:" /
       (x1-x3) (= 4.) /
       (Char1-Char3) (= $5.)/;
   call missing(of x1-x3,of Char1-Char3);
   put "Values after the CALL:" /
       (x1-x3) (= 4.) /
       (Char1-Char3) (= $5.)//;
datalines;
1 2 3 a b c
100 . 200 One Two Three
;
```

As you can see from the following output, this CALL routine sets the value of all of its arguments to a SAS missing value.

```
Values before the CALL:
x1=1  x2=2  x3=3
Char1=a Char2=b Char3=c

Values after the CALL:
x1=.  x2=.  x3=.
Char1=  Char2=  Char3=


Values before the CALL:
x1=100 x2=.  x3=200
Char1=One Char2=Two Char3=Three

Values after the CALL:
x1=.  x2=.  x3=.
Char1=  Char2=  Char3=
```

C h a p t e r   1 1

State and ZIP Code Functions

Introduction

This group of functions allows you to convert between FIPS (Federal Information Processing Standards) codes, ZIP codes, two-digit state postal codes, and state names and abbreviations. For example, these functions let you store a ZIP code and print out a state name or abbreviation, computed from the code.

Two functions added with SAS 9.2, GEODIST and ZIPCITYDISTANCE, compute distances between two sets of longitude and latitude or between two ZIP codes, respectively.

Functions That Convert FIPS Codes

This group of functions takes a FIPS code as an argument and produces either an uppercase state name, a mixed-case state name, or a two-character standard state abbreviation.

Function: **FIPNAME**

Purpose: To convert a FIPS code to an uppercase state name.

Syntax: FIPNAME (*FIPS-code*)

FIPS-code is a numeric variable or an expression that represents a standard FIPS code.

Examples

For these examples, FIPS1 = 27, FIPS2 = 2.

Function	Returns
FIPNAME(FIPS1)	MINNESOTA
FIPNAME(FIPS2)	ALASKA
FIPNAME(34)	NEW JERSEY
FIPNAME(999)	Error written to the SAS log

See Program 11.1 for a sample program.

Function: **FIPNAMEL**

Purpose: To convert a FIPS code to a mixed-case state name.

Syntax: FIPNAMEL (*FIPS-code*)

FIPS-code is a numeric variable or an expression that represents a standard FIPS code.

Examples

For these examples, FIPS1 = 27, FIPS2 = 2.

Function	Returns
FIPNAMEL(FIPS1)	Minnesota
FIPNAMEL(FIPS2)	Alaska
FIPNAMEL(34)	New Jersey
FIPNAMEL(999)	Error written to the SAS log

See Program 11.1 for a sample program.

Function: **FIPSTATE**

Purpose: To convert a FIPS code to a two-character state code.

Syntax: **FIPSTATE (*FIPS-code*)**

FIPS-code is a numeric variable or an expression that represents a standard FIPS code.

Examples

For these examples, FIPS1 = 27, FIPS2 = 2.

Function	Returns
FIPSTATE(FIPS1)	MN
FIPSTATE(FIPS2)	AL
FIPSTATE(34)	NJ
FIPSTATE(999)	Error written to the SAS log

Program 11.1: Converting FIPS codes to state names and abbreviations

```
***Primary functions: FIPNAME, FIPNAMEL, and FIPSTATE;

data fips;
   input Fips @@;
   Upper_state = fipname(Fips);
   Mixed_state = fipnamel(Fips);
   Abbrev      = fipstate(Fips);
datalines;
1 2 3 4 5 34 . 50 95 99
;
title "Listing of Data Set FIPS";
proc print data=fips noobs;
run;
```

Explanation

This straightforward program converts each of the FIPS codes to an uppercase and a mixed-case state name as well as a two-character state abbreviation. When the argument for these functions is not a valid FIPS code (which range from 1 to 95, as of this writing), the state name is "INVALID CODE" (in either uppercase or mixed-case) or two dashes for the state abbreviation (for example, the FIPS code 3). FIPS codes resulting in locations outside the U.S. result in territory names as values and values of – – for the state abbreviations (for example, FIPS code 95). A missing value produces blanks for all three values. Values greater than the number of valid FIPS codes cause an error message to be printed to the log, and missing values result from all three functions. You can verify this by inspecting the following listing:

```
Listing of Data Set FIPS

Fips    Upper_state      Mixed_state      Abbrev

 1      ALABAMA          Alabama          AL
 2      ALASKA           Alaska           AK
 3      INVALID CODE     Invalid Code     --
 4      ARIZONA          Arizona          AZ
 5      ARKANSAS         Arkansas         AR
34      NEW JERSEY       New Jersey       NJ
 .
50      VERMONT          Vermont          VT
95      PALMYRA ATOLL    Palmyra Atoll    --
99
```

Functions That Convert State Codes

This set of functions all take two-character state codes and return FIPS codes or state names.

Function: STFIPS

Purpose: To convert a two-character state code (e.g., 'NJ') to a FIPS code.

Syntax: STFIPS(*state-code*)

state-code is a two-character standard state abbreviation. This can be a SAS character variable, an expression, or a character constant, and can be in uppercase or lowercase.

Examples

For these examples, STATE1 = 'NJ' and STATE2 = 'nc'.

Function	Returns
STFIPS(STATE1)	34
STFIPS(STATE2)	37
STFIPS('TX')	48
STFIPS('XX')	missing value

See Program 11.2 for a sample program.

Function: STNAME

Purpose: Takes a two-character state code (e.g., 'NJ') and returns a state name in uppercase letters.

Syntax: STNAME(*state-code*)

state-code is a two-character standard state abbreviation. This can be a SAS character variable, an expression, or a character constant, and can be in uppercase or lowercase.

Examples

For these examples, STATE1 = 'NJ' and STATE2 = 'nc'.

Function	Returns
STNAME(STATE1)	NEW JERSEY
STNAME(STATE2)	NORTH CAROLINA
STNAME('TX')	TEXAS

See Program 11.2 for a sample program.

Function: **STNAMEL**

Purpose: To convert a two-character state code (e.g., 'NJ') to a state name in mixed case.

Syntax: STNAME (*state-code*)

state-code is a two-character standard state abbreviation. This can be a SAS character variable, an expression, or a character constant, and can be in uppercase or lowercase.

Examples

For these examples, STATE1 = 'NJ' and STATE2 = 'nc'.

Function	Returns
STNAME(STATE1)	New Jersey
STNAME(STATE2)	North Carolina
STNAME('TX')	Texas

Program 11.2: Converting state abbreviations to FIPS codes, and state names

---

```
***Primary functions: STFIPS, STNAME, and STNAMEL;

data state_to_other;
   input State : $2. @@;
   Fips = stfips(State);
   Upper_name = stname(State);
   Mixed_name = stnamel(State);
datalines;
NY NJ nj NC AL
;
title "Listing of Data Set STATE_TO_OTHER";
proc print data=state_to_other noobs;
run;
```

Explanation

This program uses the three ST functions to take a two-letter state abbreviation (in uppercase or lowercase) and output a FIPS code or a state name. See the following listing:

```
Listing of Data Set STATE_TO_OTHER

State   Fips   Upper_name     Mixed_name

NY       36    NEW YORK       New York
NJ       34    NEW JERSEY     New Jersey
nj       34    NEW JERSEY     New Jersey
NC       37    NORTH CAROLINA North Carolina
AL        1    ALABAMA        Alabama
```

Functions That Convert ZIP Codes

These four functions take a character or numeric ZIP code and return FIPS codes, state names (uppercase or mixed case), and state abbreviations.

Function: **ZIPFIPS**

Purpose: To convert a ZIP code (character or numeric) to a two-digit FIPS code (numeric).

Syntax: `ZIPFIPS(zip-code)`

zip-code is a numeric variable, expression, or constant (for a ZIP code starting with 0, the leading 0 is not necessary) or a character variable, expression, or constant.

Examples

For these examples, `ZIPN = 12345`, `ZIPC = '08822'`.

Function	Returns
ZIPFIPS(ZIPN)	36
ZIPFIPS(ZIPC)	34
ZIPFIPS(2*12222)	51

See Program 11.3 for a sample program.

Function: **ZIPNAME**

Purpose: To convert a ZIP code (character or numeric) to a state name (in uppercase).

Syntax: `ZIPNAME(zip-code)`

zip-code is a numeric variable, expression, or constant (for a ZIP code starting with 0, the leading 0 is not necessary) or a character variable, expression, or constant.

Examples

For these examples, ZIPN = 12345, ZIPC = '08822'.

Function	Returns
ZIPNAME(ZIPN)	"NEW YORK"
ZIPNAME(ZIPC)	"NEW JERSEY"
ZIPNAME(2*12222)	"VIRGINIA"

See Program 11.3 for a sample program.

Function: ZIPNAMEL

Purpose: To convert a ZIP code (character or numeric) to a state name (in mixed case).

Syntax: ZIPNAMEL(*zip-code*)

zip-code is a numeric variable, expression, or constant (for a ZIP code starting with 0, the leading 0 is not necessary) or a character variable, expression, or constant.

Examples

For these examples, ZIPN = 12345, ZIPC = '08822'.

Function	Returns
ZIPNAMEL(ZIPN)	"New York"
ZIPNAMEL(ZIPC)	"New Jersey"
ZIPNAMEL(2*12222)	"Virginia"

See Program 11.3 for a sample program.

Function: ZIPSTATE

Purpose: To convert a ZIP code (character or numeric) to a standard two-character state abbreviation.

Syntax: `ZIPSTATE(zip-code)`

`zip-code` is a numeric variable, expression, or constant (for a ZIP code starting with 0, the leading 0 is not necessary) or a character variable, expression, or constant.

Examples

For these examples, `ZIPN = 12345`, `ZIPC = '08822'`.

Function	Returns
ZIPSTATE(ZIPN)	"NY"
ZIPSTATE(ZIPC)	"NJ"
ZIPSTATE(2*12222)	"VA"

Program 11.3: Converting ZIP codes to FIPS codes, state names, and state abbreviations

```
   ***Primary functions: ZIPFIPS, ZIPNAME, ZIPNAMEL, and ZIPSTATE;

data zip_to_other;
   input Zip @@;
   Fips = zipfips(Zip);
   State_caps = zipname(Zip);
   State_mixed = Zipnamel(Zip);
   State_abbre = zipstate(Zip);
   format Zip z5.;
datalines;
1234 12345 08822 98765
;
title "Listing of Data Set ZIP_TO_OTHER";
proc print data=zip_to_other noobs;
run;
```

Explanation

This program uses a numeric value for the ZIP code (remember, these functions can also take character arguments) and returns the FIPS code, the state name (uppercase and mixed case), and standard state abbreviations. See the following listing:

```
Listing of Data Set ZIP_TO_OTHER


                                                 State_
   Zip    Fips    State_caps      State_mixed    abbre

01234    25      MASSACHUSETTS   Massachusetts    MA
12345    36      NEW YORK        New York         NY
08822    34      NEW JERSEY      New Jersey       NJ
98765    53      WASHINGTON      Washington       WA
```

Program 11.4: Adding a state abbreviation to an address containing only city and ZIP code

```
   ***Primary function: ZIPSTATE;

data address;
   input #1 Name $30.
         #2 Street $40.
         #3 City & $20. Zip;
   State = zipstate(Zip);

   file print;
   ***Create Mailing list;
   put Name /
       Street /
       City +(-1) ", " State Zip z5.//;
datalines;
Mr. James Joyce
123 Sesame Street
East Rockaway  11518
Mrs. Deborah Goldstein
87 Hampton Corner Road
Flemington  08822
;
```

Explanation

This program reads in three lines of an address. Note that in the third line, only a city and ZIP code are entered. The program uses the ZIPSTATE function to obtain the two-character state abbreviation and insert it into the address. Note the use of the ampersand informat modifier on the third line of the INPUT statement. This causes the delimiter to be two or more blanks, so that city names containing blanks can be read correctly. Notice that there are at least two blanks between the city name and the ZIP code. The addresses are then sent to the output device, using PUT statements (see the following listing).

```
Listing of Data Set ZIP_TO_OTHER
Mr. James Joyce
123 Sesame Street
East Rockaway, NY 11518


Mrs. Deborah Goldstein
87 Hampton Corner Road
Flemington, NJ 08822
```

Functions That Compute Geographical Distance

Function: **GEODIST**

Purpose: To compute the straight line distance ("as the crow flies") between two sets of latitude and longitude coordinates. By default, the distance is computed in kilometers. If you want your value to be in miles, be sure to use the m option.

Syntax: GEODIST(*latitude-1, longitude-1, latitude-2, longitude-2 <,options>*)

latitude is the distance north or south of the equator in degrees or radians (default is degrees). Locations north of the equator have positive values; locations south of the equator have negative values. Values for *latitude* must be between 0 and 90 or 0 and –90 (in degrees) or between $-\pi$ and $+\pi$ (in radians).

longitude is the distance east or west of the prime meridian. Locations east of the prime meridian have positive values; locations to the west of the prime meridian have negative values. Value for *longitude* must be between 0 and 180 or 0 and -180 (in degrees) or between -2 π and + 2π (radians).

Options

m	specifies distance in miles.
k	specifies distance in kilometers (default).
d	specifies that values of latitude and longitude are in degrees (default).
r	specifies that values of latitude and longitude are in radians.

Examples

For these examples, LAT1 = 40, LONG1 = -74, LAT2 = 30, LONG2 = -99.

Function	Returns
GEODIST(LAT1,LONG1,LAT2,LONG2,'m')	1569.71 (rounded)
GEODIST(LAT1,LONG1,LAT2,LONG2)	2525.67 (rounded)
GEODIST(45,-90,46,-91,'md')	84.44 (rounded)

Function: ZIPCITYDISTANCE

Purpose: Computes the straight line distance (in miles) between two ZIP codes. This function uses the centroid of the ZIP codes for the calculation.

Syntax: ZIPCITYDISTANCE(*zip-1*, *zip-2*);

zip is a five-digit US ZIP code.

Note: In order to use this function, you must have a file containing the coordinates of the U.S. ZIP codes. You can check if you have this file by running the following:

```
proc contents data=SASHELP.ZIPCODE;
run;
```

If you don't have the ZIP code file or want to be sure you have the latest version of the file, you can download a transport file containing the ZIP code information from the following Web site:

http://support.sas.com/rnd/datavisualization/mapsonline/html/misc.html.

Select Zipcode Dataset from the Name column to begin the download process. You must execute the CIMPORT procedure after you download and unzip the data set.

An example of how to use PROC CIMPORT is shown here:

```
proc cimport infile='c:\zipcodes\zipcode_jan09.cpt'
   data=sashelp.zipcode;
run;
```

where zipcode_jan09.cpt was the name of the file downloaded from the SAS Web site.

Examples

For these examples, `zip1 = 08822` and `zip2 = 78010`.

Function	Returns
ZIPCITYDISTANCE(ZIP1,ZIP2)	1542.1
ZIPCITYDISTANCE(08822,78010)	1542.1
ZIPCITYDISTANCE(78010,08822)	1542.1

Program 11.5: Measuring the distance between two coordinates or two ZIP codes

```
***Primary functions: GEODIST ZIPCITYDISTANCE;
data distance;
   *Three Bridges New Jersey: N 40 31.680 W 74 47.968
    Camp Verde Texas: N 29 54.597, W 99 04.242;
   *See the explanation below to understand the calculations
    in the next four lines of code;

   ThreeBridges_N = 40 + 31.680/60;
   ThreeBridges_W = -(74 + 47.968/60);
   CampVerde_N = 29 + 54.597/60;
   CampVerde_W = -(99 + 4.242/60);
   Distance_Geo = geodist(ThreeBridges_N, ThreeBridges_W,
                          CampVerde_N, CampVerde_W, 'MD');
   Distance_Zip = zipcitydistance(08822, 78010);
run;

title "Distance in Miles between Three Bridges NJ and "
      "Camp Verde TX";
proc print data=distance noobs;
run;
```

Explanation

The latitude and longitudes in the first comment are given in degrees, minutes, and decimal minutes, which is standard for most GPS devices. Other possible coordinate systems are degrees, minutes, and seconds and UTM (used by geological survey maps and the military). There are several Web sites that will convert among these various formats.

If you supply coordinates to the GEODIST function in degrees, you need to convert the degrees, minutes, and decimal minutes into degrees. Since there are 60 minutes to a degree, you need to add the number of degrees plus 1/60 of the minutes value. As a matter of interest, probably to no one besides this author, the two locations are where the author lives in New Jersey and Texas.

Remember that the default result for the GEODIST function is in kilometers, so be sure to use the m option if you want the result in miles. The two ZIP codes for the ZIPCITYDISTANCE function are the corresponding ZIP codes for Three Bridges, New Jersey and Camp Verde, Texas. Here is the output:

Distance in Miles between Three Bridges NJ and Camp Verde TX					
Three Bridges_ N	Three Bridges_ W	Camp Verde_N	Camp Verde_W	Distance_ Geo	Distance_ Zip
40.528	-74.7995	29.9100	-99.0707	1548.93	1542.1

The small differences in distance are due to the fact that SAS uses the centroid of the ZIP codes in its calculations, and the coordinates given are the actual coordinates at each of my two houses.

C h a p t e r   1 2

Trigonometric Functions

Introduction

Rather than cover all the trigonometric functions, I've decided to demonstrate just the three basic ones and their inverses. The remaining trigonometric functions work essentially the same way and their use is left as an "exercise for the reader." The salient point to remember when dealing with the trigonometric functions is that the angles are always expressed in radians.

Three Basic Trigonometric Functions

All three functions are demonstrated in the single program that follows the syntax and examples.

Function: COS

Purpose: To compute the cosine of an angle (expressed in radians).

Syntax: COS(*angle*)

angle is the angle in radians.

Note: Multiply the angle in degrees by $\pi/180$ to convert to radians. For example, the cosine of 60 degrees (1.0472 radians) is equal to .5.

Examples

Function	Returns
COS(0)	1
COS(.5236)	.8660

Note: .5236 radians = 30 degrees.

Function: SIN

Purpose: To compute the sine of an angle (expressed in radians).

Syntax: SIN(*angle*)

angle is the angle in radians.

Note: Multiply the angle in degrees by $\pi/180$ to convert to radians. For example, the sine of 30 degrees (.5236 radians) is equal to .5.

Examples

Function	Returns
SIN(0)	0
SIN(.5236)	.5

Note: .5236 radians = 30 degrees.

Function: **TAN**

Purpose: To compute the tangent of an angle (expressed in radians).

Syntax: `TAN(angle)`

angle is the angle in radians.

Note: Multiply the angle in degrees by $\pi/180$ to convert to radians. For example, the tangent of 45 degrees (.7854 radians) is equal to 1.

Examples

Function	Returns
TAN(0)	0
TAN(.5236)	.5774

Note: .5236 radians = 30 degrees.

Program 12.1: **Creating a table of trigonometric functions**

```
***Primary functions: COS, SIN, TAN
***Other function: CONSTANT;

data trig_table;
   if _n_ = 1 then Pi = constant('pi');
   retain Pi;

   do Angle = 0 to 360 by 10;
      Radian = Pi*Angle/180;
      Sin = sin(Radian);
      Cos = cos(Radian);
      Tan = tan(Radian);
      output;
   end;

   drop Pi Radian;
run;

options ls=22 missing='-';
title "Table of Basic Trig Functions";
proc report data=trig_table nowd panels=99;
   columns angle sin cos tan;
   define angle / display 'Angle' width=5 format=4.;
```

```
   define sin / display 'Sin' width=6 format=6.4;
   define cos / display 'Cos' width=6 format=6.4;
   define tan / display 'Tan' width=6 format=6.2;
run;
```

Explanation

The CONSTANT function is a convenient way to obtain an accurate value for π. With efficiency in the back (or front) of your mind, you realize you don't want to compute this value for each iteration of the DATA step. The IF-THEN statement (IF _N_ = 1) ensures that the calculation is performed only once and the RETAIN statement ensures that the value does not get replaced by a missing value as the DATA step iterates.

In this program, to make the output smaller, I set up the trigonometric table to show the values for every 10 degrees. To make the output look nicer than the standard PROC PRINT output and to print multiple columns on a single page, I used PROC REPORT. One of my main reasons for using PROC REPORT (since I do very little fancy reporting) is to use the PANELS option. If you set the value to a large number (such as 99), PROC REPORT will fit as many panels as it can on the page, provided that the number of lines of output is greater than the pagesize value (PAGESIZE was set to 22 for this program). Here is the nice-looking table produced by this report:

```
Table of Basic Trig Functions

 Angle    Sin     Cos     Tan     Angle    Sin     Cos     Tan
    0   0.0000  1.0000   0.00      190   -.1736  -.9848   0.18
   10   0.1736  0.9848   0.18      200   -.3420  -.9397   0.36
   20   0.3420  0.9397   0.36      210   -.5000  -.8660   0.58
   30   0.5000  0.8660   0.58      220   -.6428  -.7660   0.84
   40   0.6428  0.7660   0.84      230   -.7660  -.6428   1.19
   50   0.7660  0.6428   1.19      240   -.8660  -.5000   1.73
   60   0.8660  0.5000   1.73      250   -.9397  -.3420   2.75
   70   0.9397  0.3420   2.75      260   -.9848  -.1736   5.67
   80   0.9848  0.1736   5.67      270  -1.000   -.0000     -
   90   1.0000  0.0000     -       280   -.9848  0.1736  -5.67
  100   0.9848  -.1736  -5.67      290   -.9397  0.3420  -2.75
  110   0.9397  -.3420  -2.75      300   -.8660  0.5000  -1.73
  120   0.8660  -.5000  -1.73      310   -.7660  0.6428  -1.19
  130   0.7660  -.6428  -1.19      320   -.6428  0.7660  -0.84
  140   0.6428  -.7660  -0.84      330   -.5000  0.8660  -0.58
  150   0.5000  -.8660  -0.58      340   -.3420  0.9397  -0.36
  160   0.3420  -.9397  -0.36      350   -.1736  0.9848  -0.18
  170   0.1736  -.9848  -0.18      360   -.0000  1.0000  -0.00
  180   0.0000  -1.000  -0.00
```

Three Inverse Trigonometric Functions

These functions compute the inverse of the three basic trigonometric functions. Given a value, these functions compute an angle (expressed in radians).

Function: ARCOS

Purpose: To compute the inverse cosine or arccosine. The resulting angle is in radians. To convert from radians to degrees, multiply by the fraction $(180/\pi)$. For example, the arccosine of .5 is 1.0472, which is equal to 60 degrees.

Syntax: ARCOS(*value*)

value is a numeric value between −1 and +1.

Examples

Function	Returns
ARCOS(.5)	1.0472 (60 degrees)
ARCOS(1)	0

See Program 12.2 for a sample program.

Function: ARSIN

Purpose: To compute the inverse sine or arcsine. The resulting angle is in radians. To convert from radians to degrees, multiply by the fraction $(180/\pi)$. For example, the arcsine of .5 is 0.5236, which is equal to 30 degrees.

Syntax: ARSIN(*value*)

value is a numeric value between −1 and +1.

Examples

Function	Returns
ARSIN(.5)	.5326 (30 degrees)
ARSIN(1)	1.5708 (90 degrees)

Program 12.2: Computing arccosines and arcsines

```
    ***Primary functions: ARCOS and ARSIN
    ***Other function: CONSTANT;

data arc_d_triumph;
   if _n_ = 1 then Pi = constant('pi');
   retain Pi;
   drop Pi;

   do Value = 0 to 1 by .1;
      Cos_radian = arcos(Value);
      Cos_angle = Cos_radian * 180/Pi;
      Sin_radian = arsin(Value);
      Sin_angle = Sin_radian * 180/Pi;
      output;
   end;
run;

title "Listing of data set Arc D Triumph";
proc print data=arc_d_triumph noobs;
run;
```

Explanation

For each of the values between 0 and 1, this program computes the angle in radians and degrees, corresponding to the arccosine and arcsine values. Here is the listing:

```
Listing of data set ARC_D_TRIUMPH

          Cos_       Cos_       Sin_       Sin_
Value     radian     angle      radian     angle

 0.0      1.57080    90.0000    0.00000     0.0000
 0.1      1.47063    84.2608    0.10017     5.7392
 0.2      1.36944    78.4630    0.20136    11.5370
 0.3      1.26610    72.5424    0.30469    17.4576
 0.4      1.15928    66.4218    0.41152    23.5782
 0.5      1.04720    60.0000    0.52360    30.0000
 0.6      0.92730    53.1301    0.64350    36.8699
 0.7      0.79540    45.5730    0.77540    44.4270
 0.8      0.64350    36.8699    0.92730    53.1301
 0.9      0.45103    25.8419    1.11977    64.1581
 1.0      0.00000     0.0000    1.57080    90.0000
```

Function: ATAN

Purpose: To compute the inverse tangent or arctangent. The resulting angle is in radians. To convert from radians to degrees, multiply by the fraction $(180/\pi)$. For example, the arctangent of .5 is .4646.

Syntax: `ATAN(value)`

   value is a numeric value.

Examples

Function	Returns
`ATAN(99999)`	`1.5708` (89.9999 degrees)
`ATAN(1)`	`.7854` (45 degrees)
`ATAN(10)`	`1.47113` (84.2842 degrees)

Program 12.3: Computing arctangents

```
***Primary function: ATAN;
***Other function: CONSTANT;

data on_a_tangent;
   if _n_ = 1 then Pi = constant('pi');
   retain Pi;
   drop Pi;

   do Value = 0 to 10;
      Tan_radian = atan(Value);
      Tan_angle = Tan_radian * 180/Pi;
      output;
   end;
run;

title "Listing of Data Set ON_A_TANGENT";
proc print data=on_a_tangent noobs;
run;
```

Explanation

This program is similar to Program 12.2 except that the values go from 0 to 10. Here is the output:

```
Listing of Data Set ON_A_TANGENT

           Tan_        Tan_
Value      radian      angle

   0      0.00000      0.0000
   1      0.78540     45.0000
   2      1.10715     63.4349
   3      1.24905     71.5651
   4      1.32582     75.9638
   5      1.37340     78.6901
   6      1.40565     80.5377
   7      1.42890     81.8699
   8      1.44644     82.8750
   9      1.46014     83.6598
  10      1.47113     84.2894
```

C h a p t e r   1 3

Macro Functions

Introduction

This chapter covers only those macro functions that you can use during DATA step execution. Even these functions are discussed only briefly. If you want to learn more about these functions and the extensive set of functions used in the macro language, see the many fine books and reference materials currently available. In particular, *SAS Macro Programming Made Easy, Second Edition* by Michele Burlew, *Carpenter's Complete Guide to the SAS Macro Language, Second Edition* by Art Carpenter, and *SAS 9.2 Macro Language: Reference* are highly recommended.

Two of the macro CALL routines, CALL SYMPUT and CALL SYMPUTX, are used to assign the value of a DATA step variable to an existing macro variable. They can also create a new macro variable and assign it a value during the execution of the DATA step. There are several differences between CALL SYMPUT and CALLSYMPUTX, as follows.

CALL SYMPUTX left-justifies both arguments and trims trailing blanks. CALL SYMPUT does not left-justify the arguments and trims trailing blanks from the first argument only.

CALL SYMPUTX takes numeric values and formats them with the BEST32. format before placing them into a macro variable. CALL SYMPUT uses a field width of 12 characters.

CALL SYMPUTX does not write a note to the SAS log when the second argument is numeric. CALL SYMPUT does write a note to the log stating that numeric values were converted to character values.

CALL SYMPUTX enables you to specify the symbol table in which to store the macro variable; CALL SYMPUT does not.

Function: **CALL SYMPUT**

Purpose: To assign a value to a macro variable during the execution of a DATA step.

Syntax: `CALL SYMPUT (macro-var, value)`

`macro-var` is the name of a new or existing macro variable. It is either a character literal in quotation marks, a character variable, or a character expression.

`value` is a character literal, numeric or character variable, or a DATA step expression. Note that if value is numeric, SAS will perform a numeric-to-character conversion and write a note to the SAS log.

Examples

For these examples, `CHAR = "BIGMAC"` and `STRING = "Hello"`.

Function	Returns
`CALL SYMPUT(CHAR, STRING)`	Macro variable `BIGMAC` has a value of the character string HELLO
`CALL SYMPUT("MAC", STRING)`	Macro variable `MAC` has a value of the character string HELLO
`CALL SYMPUT("MAC", "Goodbye")`	Macro variable `MAC` has a value of the character string Goodbye

Function: **CALL SYMPUTX**

Purpose: To assign a value to a macro variable during the execution of a DATA step. Leading and trailing blanks are removed from the value before it is assigned to the macro variable.

Syntax: CALL SYMPUTX (*macro-var*, *value* <,*symbol-table*>)

macro-var is the name of a new or existing macro variable. It is either a
character literal in quotation marks, a character variable, or a character
expression.

value is a character literal, numeric or character variable, or DATA step
expression.

symbol-table is a character literal, variable, or expression. The first
non-blank character in *symbol-table* specifies the symbol table in
which to store the macro variable. The following values are valid as the first
non-blank character in *symbol-table*:

G	stores the macro variable in the global symbol.
L	stores the macro variable in the most local symbol that exists.
F	stores the macro variable in the most local symbol table in which the macro variable exists, or if the macro variable does not exist, stores it in the most local symbol table.

Note: If you omit *symbol-table*, or if *symbol-table* is blank, CALL
SYMPUTX stores the macro variable in the same symbol table as does the
CALL SYMPUT routine.

Examples

For these examples, CHAR = "BIGMAC" and STRING = " Hello".

Function	Returns
CALL SYMPUTX(CHAR, STRING)	Macro variable BIGMAC has a value of the character string HELLO
CALL SYMPUTX("MAC", STRING)	Macro variable MAC has a value of the character string HELLO
CALL SYMPUTX("MAC"," Goodbye ")	Macro variable MAC has a value of the character string Goodbye

**Program 13.1: Using the SYMPUT and SYMPUTX CALL routines to assign
                 a value to a macro variable during the execution of a DATA
                 step**

```
***Primary functions: SYMPUT and SYMPUTX;

data test;
   input String $char10.;
   call symput("Notx",String);
   call symputx("Yesx",String);
datalines;
   abc
;
data _null_;
   Nostrip = ":" || "&Notx" || ":";
   Strip   = ":" || "&Yesx" || ":";
   put "Value of Nostrip is  " Nostrip;
   put "Value of Strip is  " Strip;
run;
```

Explanation

The value of STRING is three blanks, followed by the letters abc, followed by four blanks.
When CALL SYMPUT is used to assign this value to the macro variable NOTX, the leading
and trailing blanks are included in the value; when CALL SYMPUTX is used, the leading
and trailing blanks are stripped. To see that this is the case, examine the results of the DATA
NULL step. It concatenates colons to the beginning and end of the macro variable so that
the blanks can be visualized. Here are the two lines sent to the SAS log:

```
Value of Nostrip is  :   abc    :
Value of Strip is  :abc:
```

**Program 13.2: Passing DATA step values from one step to another, using
                 macro variables created by CALL SYMPUT and CALL
                 SYMPUTX**

```
***Primary functions: SYMPUT and SYMPUTX
***Other functions: STRIP and PUT;

data sum;
   infile "c:\books\functions\data.dta" end=last;
   input n @@;
   sum + n;
   count + 1;
   if last then do;
```

```
      call symput("Sum_of_n",strip(put(Sum,3.)));
      call symputx("Number",put(Count,3.));
   end;
run;

title "Listing of Data Set NEXT";
title2 "Summary data: There were &NUMBER values";
title3 "The sum was &SUM_OF_N";
proc print data=sum noobs;
   var n;
run;
```

Explanation

In this example, you want to put summary information computed in a DATA step in the titles of the PROC PRINT step that follows. Both SYMPUT and SYMPUTX are used in order to demonstrate the difference between them. In the first call to SYMPUT, the STRIP function was needed to strip leading and trailing blanks from the value SUM, before assigning it to the macro variable SUM_OF_N. The next call to SYMPUTX shows the advantage of using this CALL routine: the macro variable NUMBER does not contain any leading or trailing blanks. The output from PROC PRINT is shown next:

```
Listing of Data Set NEXT
Summary data: There were 6 values
The sum was 29

n

1
6
3
4
8
7
```

Function: RESOLVE

RESOLVE executes a macro expression. The expression can be a macro variable or a macro itself. If the expression is a macro, the text generated by the macro is returned.

There are some similarities between the RESOLVE and SYMGET functions. SYMGET gets the current value of a macro variable during execution. If the value of the macro variable is modified during execution with SYMPUT, and then SYMGET is called, the new value is retrieved.

Both functions are dynamic in nature in that they both interact with the macro during execution of a DATA step.

This is a complicated topic that is beyond the scope of this book. If you want to learn more about the differences between SYMGET and RESOLVE, see the *SAS 9.2 Macro Language: Reference* in the Knowledge Base, available at http://support.sas.com/documentation.

Purpose: To return a value from a text expression to a DATA step variable during DATA step execution.

Syntax: RESOLVE(*character-value*)

    character-value is a character literal in single quotation marks, the name of a SAS character variable, or a character expression that the DATA step resolves to a macro text expression or a SAS statement.

Examples

For these examples, CHAR = '&MAC' and X=3. Note the use of single quotation marks here so that the macro processor does not resolve the argument. The value of &MAC is "Hello", and the value of &MAC3 is "Goodbye".

Function	Returns
RESOLVE(CHAR)	"Hello"
RESOLVE('&MAC' || LEFT(PUT(X,3.)))	"Goodbye"

Program 13.3: Using RESOLVE to pass DATA step values to macro variables during the execution of the DATA step

```
***Primary function: RESOLVE;

%let x1 = 10;
%let x2 = 100;
%let x3 = 1000;
data test;
   input n @@;
```

```
    Value = resolve('&x' || left(put(n,3.)));
    put _all_;
datalines;
1 3 2 1
;
```

Explanation

As the DATA step executes, the first value of N (1) is concatenated with the value '&X' and the result is the value '&X1', which in turn is resolved to the value of the macro variable X1, which is 10. The subsequent iterations of the DATA step assign different values for the argument of the RESOLVE function. The result is the various values for the variable VALUE as shown in the following four lines from the SAS log.

```
n=1 Value=10 _ERROR_=0 _N_=1
n=3 Value=1000 _ERROR_=0 _N_=2
n=2 Value=100 _ERROR_=0 _N_=3
n=1 Value=10 _ERROR_=0 _N_=4
```

Function: CALL EXECUTE

Purpose: This function allows you to execute a macro from within a DATA step. For example, based on data values of DATA step variables, you can conditionally execute macros. Since CALL EXECUTE executes during the execution phase of the DATA step and since macro values are needed prior to DATA step execution, see the documentation in the Knowledge Base at http://support.sas.com/documentation or one of the books devoted specifically to macro programming for tips before you start.

The SAS code generated by CALL EXECUTE executes when the DATA step completes. That is, if you use CALL EXECUTE to generate a PROC invocation, the PROC does not execute until the DATA step completes.

Syntax: `CALL EXECUTE (character-value)`

character-value is a character literal in quotation marks, the name of a SAS character variable, or a character expression that the DATA step resolves to a macro text expression or a SAS statement.

Examples

For these examples, NAME = "%MYMACRO".

Function	Result
CALL EXECUTE ('%MYMACRO')	Executes a macro called MYMACRO
CALL EXECUTE(NAME)	Executes a macro called MYMACRO
CALL EXECUTE('%MAC(DSN)')	Executes a macro called MAC and passes the argument DSN to the macro

Program 13.4: Using CALL EXECUTE to conditionally execute a macro

```
***Primary function: CALL EXECUTE;

%macro simple(Dsn=);
   title "Simple listing - today is &sysday";
   proc print data=&Dsn noobs;
   run;
%mend simple;

%macro complex(Dsn=);
   proc means data=&Dsn n mean std clm maxdec=2;
      title "Complex statistics - Friday";
   run;
%mend complex;

data test;
   input x y @@;
datalines;
7 5 1 2 3 4 9 8
;
data _null_;
   if "&Sysday" ne "Friday" then call execute('%simple(Dsn=Test)');
   else call execute('%Complex(Dsn=Test)');
run;
```

Explanation

The two macros SIMPLE and COMPLEX produce either a simple PROC PRINT listing or a PROC MEANS summary of data, respectively. In this example, the data set TEST contains two variables, X and Y, and the data set is used in the next DATA _NULL_ step. Finally, the program tests the value of the automatic macro variable, &SYSDAY. If the value is not "Friday", the SIMPLE macro executes; if the value of &SYSDAY is "Friday", the COMPLEX macro executes. The following listing was produced when this program was run on a Tuesday:

```
Simple listing - today is Tuesday

x    y

7    5
1    2
3    4
9    8
```

Running this program on a Friday results in the following listing:

```
Complex statistics - Friday

The MEANS Procedure

                                        Lower 95%      Upper 95%
Variable   N        Mean     Std Dev    CL for Mean    CL for Mean

x          4        5.00     3.65        -0.81          10.81
y          4        4.75     2.50         0.77           8.73
```

Program 13.5: A non-macro example of CALL EXECUTE

```
    ***Primary function: CALL EXECUTE;

data execute;
   String = "proc print data=execute noobs;
   title 'Listing of data set EXECUTE'; run;";
   drop String;
   infile 'c:\books\functions\data2.dat' end=Last;
   input x;
   if Last then call execute(String);
run;
```

Explanation

This program demonstrates how CALL EXECUTE can work with any SAS code. In this example, CALL EXECUTE executes the code stored in STRING after the DATA step has read in the last data value. The result of running this program is simply the output from the PROC PRINT as shown here:

```
Listing of data set EXECUTE

x

2
3
4
7
6
5
```

Function: **SYMGET**

Purpose: To obtain the value of a macro variable during DATA step execution.

Syntax: SYMGET(*character-value*)

character-value is a character literal in quotation marks, the name of a SAS character variable, or a character expression that the DATA step resolves to the name of a macro variable.

Note: You do not include the ampersand (&) as part of this argument.

Examples

For these examples, the statement %LET M = Monday was submitted in open code. Also, the statement CHARVAR = "M"; was part of the DATA step.

Function	Returns
SYMGET('M')	Monday
SYMGET(CHARVAR)	Monday

Program 13.6: Using the SYMGET function to assign a macro value to a DATA step variable

```
   ***Primary function: SYMGET
   ***Other function: CATS;

%let x1 = 5;
%let x2 = 10;
%let x3 = 15;
data test;
   input Type $ Value @@;
   Mult = symget(cats('x',Type));
   New_value = Mult * Value;
datalines;
1 5  2 5  3 5
;
title "Listing of Data Set TEST";
proc print data=test;
run;
```

Explanation

The value of TYPE is read from the input data. The CATS function concatenates the character "X" and TYPE, and also strips leading and trailing blanks. (See Chapter 1, "Character Functions," for details.) So, in the first iteration of the DATA step, the argument for the SYMGET function is "X1" and the value of the macro variable X1 is 5. Thus, the variable MULT is assigned a value of 5 in the first observation. The following listing of the data set TEST shows that MULT takes on the values of 5, 10, and 15 in the three observations, respectively.

```
Listing of Data Set TEST

                            New_
Obs    Type    Value    Mult    value

 1      1        5        5       25
 2      2        5       10       50
 3      3        5       15       75
```

C h a p t e r   1 4

SAS File I/O Functions

Introduction

The functions in this chapter address SAS data sets. There are functions to determine the number of observations in a data set, the number of character and numeric variables, and the formats and labels associated with each variable. You can readily obtain much of the same information produced by PROC CONTENTS by using these functions.

If you need to obtain SAS variable information, the V functions are also available and may be more convenient to use. See Chapter 15, "Variable Information Functions."

Creating Test Data Sets

Two SAS data sets, TEST and MISS, are used to demonstrate many of the functions in this chapter. The following two SAS programs create these two data sets:

Program 14.1: Program to create two SAS data sets, TEST and MISS

```
data test(label="This is the data set label");
   length X3 4;
   input @1  Subj      $char3.
         @4  DOB       mmddyy10.
         @14 (X1-X3)      (1.)
         @17 F_name    $char10.
         @27 L_name    $char15.;
   label Subj   = 'Subject'
         DOB    = 'Date of Birth'
         X1     = 'The first X'
         X2     = 'Second X'
         X3     = 'Third X'
         F_name = 'First Name'
         L_name = 'Last Name';
   format DOB mmddyy10. X3 roman.;
datalines;
00110/21/1946123George    Hincappie
00211/05/1956987Pincus    Zukerman
00701/01/19903..Jose      Juarez
;

data miss;
   input X Y A$ Z1-Z3;
datalines;
1 2 abc 4 5 6
999 5 xyz 999 4 5
999 999 xxx 9 8 999
2 8 ZZZ 999 999 999
;
proc sort data=test;
   by subj;
run;
title "Listing of Data Set TEST";
proc print data=test noobs label;
run;

title "Listing of Data Set MISS";
proc print data=miss noobs;
run;
```

The listing of these two data sets, output from PROC PRINT, is shown here:

```
Listing of Data Set TEST

Third                Date of      The    Second   First
   X     Subject       Birth    first X     X      Name    Last Name

 III      001       10/21/1946     1        2     George   Hincappie
 VII      002       11/05/1956     9        8     Pincus   Zukerman
   .      007       01/01/1990     3        .     Jose     Juarez

Listing of Data Set MISS

   X      Y      A      Z1     Z2     Z3

   1      2     abc      4      5      6
 999      5     xyz    999      4      5
 999    999     xxx      9      8    999
   2      8     ZZZ    999    999    999
```

Functions That Determine if SAS Data Sets Exist, and That Open and Close Files

Function: EXIST

Purpose: To determine if a SAS data set exists. This is useful prior to opening the data set and attempting to determine data set attributes. The function returns a 1 if the data set exists and a 0 otherwise.

Syntax: EXIST(*member-name* <, *type*>)

member-name is a SAS data set name, a view, or a SAS catalog.

type is either data (default) or view.

Note: See the documentation in the Knowledge Base, available at http://support.sas.com/documentation for a list of other data types.

Examples

For these examples, TEST="MYLIB.MYDATA".

Function	Returns
EXIST('MYLIB.MYDATA')	1　(if the data set exists), 0 otherwise
EXIST(TEST,VIEW)	1　(if the data view exists), 0 otherwise

See Program 14.2 for an example of the EXIST function.

Function:　OPEN

Purpose:　To open a SAS data set and return a data set ID. An ID is necessary for many of the functions in this chapter. If the data set cannot be opened, OPEN returns a 0.

Syntax:　OPEN(*data-set-name* <,'*mode*'>)

data-set-name is a SAS data set name of the form *libref.data-set-name*.

mode is an optional argument that indicates the type of access.

I	is the default value and opens the file for input using random access if available.
IN	opens the file with sequential access, allowing you to revisit observations.
IS	opens the file with sequential access without allowing you to revisit observations.

Examples

For these examples, TEST="MYLIB.MYDATA".

Function	Result
OPEN(TEST)	Opens MYLIB.MYDATA with random access
OPEN(TEST,"IN")	Opens MYLIB.MYDATA with sequential access
OPEN('MYLIB.FRED')	Opens permanent SAS data set MYLIB.FRED with random access

See Programs 14.2 and 14.3 for examples of the OPEN function.

Function: CLOSE

Purpose: To close a SAS data set after it has been opened with the OPEN function. It is good practice to close data sets as soon as possible.

Syntax: CLOSE(*dsid*)

dsid is the data set ID returned by the OPEN function.

Examples

For this example, the statement DSID = OPEN("TEST"); was previously submitted.

Function	Returns
CLOSE(DSID)	0 (if the operation was successful)

See Programs 14.2 and 14.3 for examples of the CLOSE function.

Functions That Return SAS Data Set Characteristics

The ATTRC and ATTRN functions return information about a SAS data set. Instead of using the SAS data set name as one of the arguments in the function, a SAS data set ID (DSID) is used instead. The value of this ID is returned by the OPEN function. This value can then be used as the DSID argument required by these functions.

Function: ATTRC

Purpose: To return various pieces of information concerning a SAS data set. The "C" in the function name refers to "character." The reason for that is that the various pieces of information obtainable by this function are character values. Some of the more useful attributes returned by this function are the data set label, the engine used to access the data set, and the LIBREF of the SAS library where the data set is stored.

Syntax: ATTRC(*dsid*, '*attribute*')

dsid is the data set ID returned by the OPEN function.

attribute is one of the following, placed in quotation marks.

Note: This is not a complete list. See the documentation in the Knowledge Base, available at http://support.sas.com/documentation, for a complete list.

charset	returns a value for the character set of the machine that created the data set. Possible values are ASCII, EBCDIC, ANSI (OS/2 ANSI standard ASCII), and OEM (OS/2 OEM code format).
encyrpt	returns "YES" or "NO" depending on whether the data set was encrypted.
engine	returns the name of the SAS engine needed to access the data set.
label	returns the SAS data set label.
lib	returns the LIBREF of the SAS library where the data set is located.
sorted by	returns the names of the BY variables. If the data set is not sorted, the function returns an empty string.

Examples

For these examples, the statement DSID = OPEN(TEST); was issued.

Function	Returns
ATTRC(DSID,'LABEL')	This is the data set label
ATTRC(DSID,'ENGINE')	V9
ATTRC(DSID,'CHARSET')	ANSI
ATTRC(DSID,'ENCYRPT')	NO
ATTRC(DSID,'SORTEDBY')	SUBJ (if the data set was sorted)

See Program 14.3 for a sample program.

Function: **ATTRN**

Purpose: To return various pieces of information concerning a SAS data set. The "N" in the function name refers to the fact that this function returns numeric information concerning a SAS data set. Some of the more useful attributes returned by this function include whether there are any observations or variables in the data set, the number of observations and the number of variables, and whether the data set is password-protected.

Syntax: ATTRN(*dsid*, '*attribute*')

dsid is the data set ID returned by the OPEN function.

attribute is one of the following, placed in quotation marks.

Note: This is not a complete list. See the documentation in the Knowledge Base, available at http://support.sas.com/documentation, for a complete list.

any	indicates if the data set has observations and/or variables.
-1	indicates that there are no observations and no variables.
0	indicates that there are no observations.
1	indicates that the data set has both observations and variables.
crdte	is the creation date of the SAS data set as a date-time variable.
modte	is the date the data set was last modified (also a date-time value).
nlobs	is the number of logical observations (not including those marked to be deleted) and is unaffected by any WHERE clause that may be active. This is the attribute you will probably use, rather than either of the two that follow.

nlobsf	is the number of logical observations (not including those marked to be deleted), including the effect of a FIRSTOBS=, OBS=, and any WHERE statement that may be in effect. Note that this option forces the system to read every observation and can consume large computer resources.
nobs	is the number of physical observations (including those marked for deletion).
Nvars	returns the number of variables in the data set.
pw	returns a 1 if the data set is password-protected, a 0 if it is not.

Examples

For these examples, the statement `DSID = OPEN(TEST);` was issued.

Function	Returns
`ATTRN(DSID,'ANY')`	1
`ATTRN(DSID,'NLOBS')`	3
`ATTRN(DSID,'NOBS')`	3
`ATTRN(DSID,'NVARS')`	7

Program 14.2: Macro to determine the number of observations in a SAS data set

```
***Primary functions: OPEN, EXIST, ATTRN, CLOSE;

%macro nobs(Dsn= /*Data set name */);
   if exist("&Dsn") then do;
      Dsid = open("&Dsn","i");
      Nobs = attrn(Dsid,"Nlobs");
   end;
   else Nobs=.;
   rc = close(Dsid);
%mend nobs;
```

Using the macro:

```
data useit;
   %nobs(Dsn=test);
   put Nobs=;
run;
```

Explanation

This short macro first checks to see that the data set exists, using the EXIST function. If it does, you open the data set and determine the number of observations, using the ATTRN function. If the data set does not exist, NOBS is set to a missing value. Finally, the CLOSE function closes the data set. You might want to create a macro variable (using CALL SYMPUT) with this value so you could conditionally execute SAS code, depending on the number of observations in a SAS data set. The following single line is written to the SAS log by this program:

```
Nobs=3
```

See Program 14.3 for an additional sample program.

Function: DSNAME

Purpose: To return the SAS data set name associated with a SAS data set ID (returned by the OPEN function).

Syntax: DSNAME (*dsid*)

dsid is the SAS data set ID returned by the OPEN function.

Examples

For this example, the statement OSCAR = OPEN ("TEST"); was previously submitted.

Function	Returns
DSNAME(OSCAR)	TEST

Program 14.3: Determining the number of observations, variables, and other characteristics of a SAS data set using SAS I/O functions

```
***Primary functions: OPEN, ATTRC, ATTRN, CLOSE;

title;
data _null_;
   Dsid = open ('test');
   Any = attrn(Dsid,'Any');
   Nlobs = attrn(Dsid,'Nlobs');
   Nvars = attrn(Dsid,'Nvars');
   Label = attrc(Dsid,'Label');
   Engine = attrc(Dsid,'Engine');
   Charset =attrc(Dsid,'Charset');
   Dsn = dsname(Dsid);
   RC = close(Dsid);
   file print;
   put "Characteristics of Data Set Test" /
      40*'-'/
      "Dsid ="     @11 Dsid    /
      "Any ="      @11 Any     /
      "Nlobs ="    @11 Nlobs   /
      "Nvars ="    @11 Nvars   /
      "Label ="    @11 Label   /
      "Engine ="   @11 Engine  /
      "Charset =" @11 Charset /
      "Dsn ="      @11 Dsn     /
      "RC ="       @11 RC;
   run;
```

Explanation

The OPEN function returns a data set ID that is used with the ATTRC and ATTRN functions to identify the data set for which you want information. The three ATTRN functions return very useful information about a SAS data set: whether there are any variables and/or observations, the number of observations, and the number of variables. The ATTRC function is used to determine other information about the SAS data set that may be important in a large system where the data sets are created on different platforms and may use different coding (such as ANSI, ASCII, or EBCDIC) for storing character variables. It is important to close the data set when you are done. Data sets opened in a DATA step are automatically closed, but it is a good habit to close any data set as soon as possible. The listing produced by this program follows:

```
Characteristics of Data Set Test
----------------------------------------
Dsid =    1
Any =     1
Nlobs =   3
Nvars =   7
Label =   This is the data set label
Engine =  V9
Charset = ANSI
Dsn =     WORK.TEST.DATA
RC =      0
```

Functions That Return Variable Information

See Chapter 15, "Variable Information Functions," for an alternative way to obtain variable information using the V functions.

Function: **VARFMT**

Purpose: To determine the format assigned to a SAS data set variable. If the resulting variable is not previously given a length, the length will be set to 200.

Syntax: **VARFMT(*dsid*, *var-number*)**

dsid is the SAS data set ID returned by the OPEN function.

var-number is a number reflecting the variable's order in the SAS data set. This number is returned by the VARNUM function. You can also inspect the output from PROC CONTENTS.

Examples

For this example, the statement DSID = OPEN("TEST"); was previously submitted.

Function	Returns
VARFMT(DSID,3)	The format for the 3rd variable in the data set with identifier DSID: MMDDYY10.
VARFMT(DSID,VARNUM(DSID,"DOB"))	The format for the variable DOB in the data set with identifier DSID: MMDDYY10.

See Program 14.4 for a sample program.

Function: **VARLABEL**

Purpose: To determine the label assigned to a SAS data set variable. If the resulting variable is not previously given a length, the length will be set to 200.

Syntax: **VARLABEL(*dsid, var-number*)**

dsid is the SAS data set ID returned by the OPEN function.

var-number is a number reflecting the variable's order in the SAS data set. This number is returned by the VARNUM function. You can also inspect the output from PROC CONTENTS.

Examples

For this example, the statement DSID = OPEN("TEST"); was previously submitted.

Function	Returns
VARLABEL(DSID,3)	The label assigned to the 3rd variable in the data set with identifier DSID: Date of Birth.
VARLABEL(DSID,VARNUM(DSID,"DOB"))	The label assigned to the variable GENDER in the data set with identifier DSID: Date of Birth.

See Program 14.4 for a sample program.

Function: **VARLEN**

Purpose: To determine the storage length for a SAS data set variable.

Syntax: **VARLEN(*dsid, var-number*)**

dsid is the SAS data set ID returned by the OPEN function.

var-number is a number reflecting the variable's order in the SAS data set. This number is returned by the VARNUM function. You can also inspect the output from PROC CONTENTS.

Examples

For this example, the statement DSID = OPEN("TEST"); was previously submitted.

Function	Returns
VARLEN(DSID,6)	The length format for the 6th variable in the data set with identifier DSID: 10.
VARLEN(DSID,VARNUM(DSID,"F_NAME"))	The length for the variable F_NAME in the data set with identifier DSID: 10.

Program 14.4: Determining the format, label, and length attributes of a variable using the VARFMT, VARLEN, and VARNUM functions

```
***Primary functions: OPEN, VARFMT, VARNUM, VARLABEL, VARLEN, CLOSE;

data _null_;
   length Format Label $ 32;
   Dsid = open("test");
   Order = varnum(Dsid,"x3");
   Format = varfmt(Dsid,Order);
   Label = varlabel(Dsid,Order);
   Length = varlen(Dsid,Order);
   RC = close(Dsid);
   put Order= Format= Label= Length=;
run;
```

Explanation

The OPEN function opens the data set TEST and assigns a data set identifier (DSID). The VARNUM function determines the order of variable X3 in the data set (1).

Note: When data set TEST was created, a LENGTH statement was placed at the top of the DATA step to assign a length for variable X3, thus making it the first variable in the data descriptor for the data set (see the following output).

The VARFMT function also returns the format (ROMAN) for variable X3. The VARLABEL function returns the label "Third X," and the VARLEN function returns a 4, the length assigned to the variable X3 by the aforementioned LENGTH statement in the DATA step. The following line is printed to the SAS log by this program:

```
Order=1 Format=ROMAN. Label=Third X Length=4
```

Function: **VARNAME**

Purpose: To return the name of a SAS variable, given the order in the SAS data set. This number is returned by the VARNUM function. You can also inspect the output from PROC CONTENTS.

Syntax: **VARNAME (*dsid, var-number*)**

dsid is the SAS data set ID returned by the OPEN function.

var-number is a number reflecting the variable's order in the SAS data set. This number is returned by the VARNUM function. You can also inspect the output from PROC CONTENTS.

Examples

For this example, the statement `DSID = OPEN("TEST");` was previously submitted.

Function	Returns
`VARNAME(DSID,3)`	The variable name of the 3rd variable in the data set with identifier DSID: DOB.

See Program 14.5 for a sample program.

Function: **VARNUM**

Purpose: To return the position of a SAS variable in a SAS data set, given the variable name. This function is basically the inverse of the VARNAME function.

Syntax: **VARNUM(*dsid, varname*)**

dsid is the SAS data set ID returned by the OPEN function.

varname is a name of a variable in the SAS data set with the identifier *dsid.*

Examples

For this example, the statement `DSID = OPEN("TEST");` was previously submitted, and `V = "DOB".`

Function	Returns
`VARNUM(DSID,"DOB")`	The position of a variable called "DOB" in the SAS data set with identifier DSID: 3.
`VARNUM(DSID,V)`	The position of a variable called "DOB" in the SAS data set with identifier DSID: 3.

Program 14.5: Determining a variable name, given its position in a SAS data set and vice versa

```
***Primary functions: OPEN, VARNAME, VARNUM, CLOSE;

data _null_;
   ID = open("test");
   Var_name = varname(ID,1);
   Var_pos  = varnum(ID,"DOB");
   RC = close(ID);
   put "The name of the 1st variable in data set test is: "  Var_name /
       "The position of variable DOB is: " Var_pos;
run;
```

Explanation

The data set TEST is opened with the OPEN function, and the data set identifier is assigned to the variable called ID. The VARNAME function returns the variable name assigned to the first variable in the data set TEST, and the VARNUM function returns the position of the variable DOB in the data set TEST. The following two lines are written to the SAS log:

```
The name of the 1st variable in data set test is: X3
The position of variable DOB is: 3
```

Function: VARTYPE

Purpose: To determine if a SAS data set variable is character or numeric. For character variables, the function returns a "C" and for numeric variables it returns an "N".

Syntax: VARTYPE(*dsid, var-number*)

dsid is the SAS data set ID returned by the OPEN function.

var-number is a number reflecting the variable's order in the SAS data set. This number is returned by the VARNUM function. You can also inspect the output from PROC CONTENTS.

Examples

For this example, the statement `DSID = OPEN("TEST");` was previously submitted.

Function		Returns
VARTYPE(DSID,1)	(variable X3)	"N"
VARTYPE(DSID,6)	(variable F_NAME)	"C"
VARTYPE(DSID,3)	(variable DOB)	"N"

Program 14.6: A general-purpose macro to display data set attributes such as number of observations, number of variables, variable list, variable type, formats, labels, etc.

```
***Primary functions: EXIST, OPEN, ATTRN, ATTRC, VARNAME, VARTYPE,
                       VARLEN, VARFMT, VARLABEL, and CLOSE;

%macro dsn_info(Dsn= /* Data set name */);
   title;
   data _null_;
      file print;
      if not exist("&dsn") then do;
         put "Data set &Dsn does not exist";
         stop;
      end;

      Dsid    = open("&Dsn","i");
      Nvars   = attrn(Dsid,"Nvars");
      Nobs    = attrn(Dsid,"Nlobs");
      Charset = attrc(Dsid,"Charset");
      Engine  = attrc(Dsid,"Engine");
      Encrypt = attrc(Dsid,"Encrypt");
      Label   = attrc(Dsid,"Label");
      Sort    = attrc(Dsid,"Sortedby");

      put "Information for Data Set &Dsn" /
          72*'-' // ;
      if not missing(Label) then put "Data set Label is: " Label;
      put "Data set created with engine: " Engine /
          "Character set used: " Charset;
      if encrypt = "YES" then put "Data set is encrypted";
      else put "Data set is not encrypted";
      if Sort = " " then put "Data set is not sorted";
      else put "Data set is sorted by: " Sort /;
      put
          "Number of Observations: " Nobs /
          "Number of Variables   : " Nvars /
```

```
        72*'-' /
        "***** Variable Information *****" //
        @1   "Variable Name"
        @20  "Type"
        @26  "Length"
        @34  "Format"
        @47  "Label"/
        72*'-';
    do i = 1 TO Nvars;
        Name = varname(Dsid,i);
        Type = vartype(Dsid,i);
        if Type = "C" then Type = "Char";
        else if Type = "N" then Type = "Num";

        Length = varlen(Dsid,i);
        Fmt = varfmt(Dsid,i);
        Label = varlabel(Dsid,i);

        put @1    Name
            @20   Type
            @26   Length
            @34   Fmt
            @47   Label;
    end;

    RC = close(Dsid);
  run;
%mend dsn_info;
```

Explanation

This rather long macro uses most of the functions described in this chapter to produce a very compact equivalent of PROC CONTENTS. You may want to customize it to produce just the information you find useful. The VARTYPE function determines if a variable is character or numeric.

Calling the macro %dsn_info(Dsn=Test) results in the following output:

```
Information for Data Set test
- - - - - - - - - - - - - - - - - - - - - - - - - - - - - - - - - - - - - - - - - - - - - - - - - - - - - - - - -


Data set Label is: This is the data set label
Data set created with engine: V9
Character set used: ANSI
Data set is not encrypted
Data set is sorted by: Subj

Number of Observations: 3
Number of Variables   : 7
- - - - - - - - - - - - - - - - - - - - - - - - - - - - - - - - - - - - - - - - - - - - - - - - - - - - - - - - -
***** Variable Information *****

Variable Name       Type  Length  Format       Label
- - - - - - - - - - - - - - - - - - - - - - - - - - - - - - - - - - - - - - - - - - - - - - - - - - - - - - - - -
X3                  Num   4       ROMAN.       Third X
Subj                Char  3                    Subject
DOB                 Num   8       MMDDYY10.    Date of Birth
X1                  Num   8                    The first X
X2                  Num   8                    Second X
F_name              Char  10                   First Name
L_name              Char  15                   Last Name
```

C h a p t e r   1 5

Variable Information Functions

Introduction

The functions in this chapter all return information about SAS variables. For example, VNAME returns the name of a variable, given an array reference. VTYPE is used to determine if a variable is character or numeric. The advantage of these functions over the VAR functions listed in the previous chapter is that the OPEN function is not needed to obtain variable information.

As a general class, there are Vname functions and VnameX functions. For example, there is a function called VTYPE and another function called VTYPEX. The difference is that the Vname functions need a variable name as the argument; the VnameX functions can evaluate an expression to determine the variable name.

I will not discuss the VnameX functions individually. A single example is shown for VTYPE and VTYPEX. Further, not all the V functions are discussed here. For information on some of the less frequently used functions, see *SAS 9.2 Language Reference: Concepts* in the Knowledge Base, available at http://support.sas.com/documentation. Note that the wording of arguments in this book might differ from the wording of arguments in the documentation.

Functions That Determine SAS Variable Information

Function: VTYPE

Purpose: To determine if a variable is character or numeric.

Syntax: VTYPE (*variable*)

variable is the name of a SAS variable.

Note: By default, the result of this function is a character variable of length 1. Values of "C" or "N" are returned for character and numeric variables, respectively.

Examples

Function	Returns
VTYPE (NUM) where NUM is a numeric variable	"N"
VTYPE (CHAR) where CHAR is a character variable	"C"

See Program 15.2 for an example.

Function: VTYPEX

Purpose: To determine if a variable is character or numeric.

Syntax: VTYPEX(*expression*)

expression is an expression that evaluates to a SAS variable name.

Note: By default, the result of this function is a character variable of length 1. Values of "C" or "N" are returned for character and numeric variables, respectively.

Examples

Function	Returns
VTYPEX('X' || '1') where X1 is a numeric variable	"N"
VTYPEX('CH' || 'AR') where CHAR is a character variable	"C"

Program 15.1: Creating a test data set for use with most of the V functions

```
proc format;
   value yesno 1='yes' 0='no';
   invalue read 0-5 = 777
                6-9 = 888
                other = 999;
run;
data var;
   informat Char $char1. y read2. z 3.2 Money dollar5.;
   input @1  x       1.
         @2  Char    $char1.
         @3  y       read2.
         @5  z       3.2
         @8  Money   dollar5.
         @13 (a1-a3) (1.)
         @16 Date    mmddyy10.;
   format x yesno. Money dollar8.2 z 7.4;
   label x = 'The x variable'
         y = 'Pain scale'
         Money = 'Housing cost';
   array oscar[3] a1-a3;
datalines;
1B 31231,765987
0N 86549,123234
;
```

Program 15.2: Determining a variable's type (numeric or character) using VTYPE and VTYPEX

```
***Primary functions: VTYPE, VTYPEX;

data char_num;
   set var;
   Type_x = vtype(x);
   Type_char = vtype(Char);
   Type_a1 = vtypex('a' || put(3-2,1.));
run;

title "Listing of Data Set CHAR_NUM";
proc print data=char_num noobs;
run;
```

Explanation

Since variables X and A1 are both numeric, the VTYPE and VTYPEX functions return an "N". Note that the expression for TYPE_A1 evaluates to the value "A1", which is the name of a numeric variable. This same expression could not have been used as the argument of VTYPE. TYPE_CHAR is a character variable and the VTYPE function returns a "C". Note that you do not have to predefine the length of the resulting character variables—they are set to 1 as the default length. Here is a listing of output:

```
Listing of Data Set CHAR_NUM

                                            Type_
Char  y        z      Money    x  a1 a2 a3 Date Type_x char  Type_a1

B     777  1.2300  $1765.00 yes  9  8  7   .    N      C     N
N     888  6.5400  $9123.00 no   2  3  4   .    N      C     N
```

Function: VLENGTH

Purpose: To return the storage length (determined at compile time) for a variable. See the LENGTHM function in Chapter 1, "Character Functions", for a function with a similar purpose.

Syntax: VLENGTH(*variable*)

variable is the name of a SAS variable.

Example

The variables in these examples are taken from Program 15.1.

Function	Returns
VLENGTH(CHAR)	1
VLENGTH(Y)	8
VLENGTH(MONEY)	8

Program 15.3: Determining the storage length of character and numeric variables

```
***Primary function: VLENGTH;

data _null_;
   length y 4 Name $ 20;
   x = 123;
   y = 123;
   Char = 'abc';
   Long = 'this is a long character variable';
   Pad = 'a    '; ***a followed by 4 blanks;
   Name = 'Frank';

   L_x = vlength(x);
   L_y = vlength(y);
   L_char = vlength(Char);
   L_long = vlength(Long);
   L_pad = vlength(Pad);
   L_name = vlength(Name);

   put L_x = /
       L_y = /
       L_char = /
       L_long = /
       L_pad = /
       L_name = ;
run;
```

Explanation

The VLENGTH function is used to determine the storage length for the variables in the SAS data set. A portion of the output to the SAS log is shown next. Notice in particular that the length of PAD (5) is the storage length, not the value returned by the LENGTH function.

```
L_x=8
L_y=4
L_char=3
L_long=33
L_pad=5
L_name=20
```

Function: **VNAME**

Purpose: To return the name of a SAS variable. It is most useful when you want to determine which variable a specific array element is referencing. Note there is also a CALL routine with the same name that accomplishes the same task.

Syntax: **VNAME (*variable-reference*)**

variable-reference is usually an array element. You should use a LENGTH statement to hold the result of this function. Otherwise, the result will be a character variable of length 200.

Examples

For these examples, the statement ARRAY NUMS[*] X2-X3 Y Z; was submitted previously.

Function	Returns
VNAME(NUM[4])	"Y"
VNAME(NUM[2])	"X2"

Program 15.4:   Program to replace all numeric values of 999 with a SAS missing value and to provide a list of variable names where the changes were made

```
***Primary functions: VNAME, SYMPUTX
***Other functions: DIM, MISSING;
***This data step determines the number of numeric
   variables in the data set and assigns it to a macro variable (N);
```

```
data _null_;
   set miss;
   array nums[*] _numeric_;
   call symputx('n',dim(nums));
   stop;
run;

data _null_;
   set miss end=Last;
   file print;
   array nums[*] _numeric_;
   length Names1-Names&n $ 32;
   array names[&n];
   array howmany[&n];
   retain Names1-Names&n;
   do i = 1 to dim(nums);
      if nums[i] = 999 then do;
         nums[i] = .;
         names[i] = vname(nums[i]);
         howmany[i] + 1;
      end;
   end;
   if Last then do i = 1 to &n;
      if not missing(Names[i]) then
      put howmany[i] "Values of 999 converted to missing for variable "
         Names[i];
   end;
run;
```

Explanation

The first DATA _NULL_ DATA step uses the special name _NUMERIC_ to refer to all numeric variables in the MISS data set. The DIM function determines this number, and the call to SYMPUTX assigns this value to the macro variable N.

The next DATA step sets up one array, NUMS, to represent the 'N' numeric variables and another array, NAMES, to hold the variable names of these variables. Notice that the length of NAMES is set directly in the ARRAY statement (32) so that a LENGTH statement is not necessary. As the DO loop executes, each of the numeric variables are tested to see if a value of 999 is found, and if so, the value is converted to a SAS missing value and a counter (HOWMANY[I]) is incremented to keep track of how many replacements were made. A listing produced by this program follows:

```
Listing of Data Set CHAR_NUM
2 Values of 999 converted to missing for variable X
1 Values of 999 converted to missing for variable Y
2 Values of 999 converted to missing for variable Z1
1 Values of 999 converted to missing for variable Z2
2 Values of 999 converted to missing for variable Z3
```

Program 15.5: Writing a general-purpose macro to replace all occurrences of a particular value (such as 999) to a SAS missing value and produce a report showing the number of replacements for each variable

```
;***Primary functions: VNAME, CALL SYMPUT
***Other functions: DIM, LEFT, PUT, MISSING;

%macro replace_missing (
                        Dsn=,        /* the sas data set name */
                        Miss_value= /* the value of the missing value */
                        );
    title "Report on data set &Dsn";
    data _null_;
       set &Dsn;
       array nums[*] _numeric_;
       call symput('n',dim(nums));
       stop;
    run;
    data _null_;
       set &Dsn end=Last;
       file print;
       array nums[*] _numeric_;
       array names[&n] $ 32 _temporary_ ;
       array howmany[&n] _temporary_;
       do i = 1 to dim(nums);
          if nums[i] = &Miss_value then do;
             nums[i] = .;
             Names[i] = vname(nums[i]);
             howmany[i] + 1;
          end;
       end;
       if Last then do i = 1 to &n;
          if not missing(Names[i]) then
          put howmany[i]
"Values of &Miss_value converted to missing for variable " names[i];
       end;
    run;
%mend replace_missing;
```

Explanation

This macro is basically identical to the previous program. The only difference is that the data set name and the value to be replaced are calling arguments to the macro. Just to make sure it works, I called the macro like this (Note: data set MISS was created in Program 14.1):

```
%replace_missing(Dsn=miss,Miss_value=999)
```

and produced output identical to the previous listing.

Function: VLABEL

Purpose: To return the label associated with a SAS variable. If no label is associated with the variable, the function returns the variable name. If no length is assigned to the variable holding the label, it is given a length of 200 by default.

Syntax: VLABEL(*variable*)

variable is the name of a SAS variable. VLABEL does not accept an expression for the variable name. Use VLABELX for that purpose.

Examples

For these examples, VAR1 = 'X' and VAR2 = 'Z'.

Function	Returns
VLABEL(VAR1)	"The X variable"
VLABEL(VAR2)	"Z"
VLABEL('X')	"The X variable"

See Program 15.8 for an example.

Function: CALL VNEXT

Purpose: To determine the name, type, and length of one or more DATA step variables. Each successive call to the routine returns information on the next variable in the SAS data set.

Syntax: CALL VNEXT (*name* <, *type* <,*length*>>)

name is the name of the variable that will hold the variable name. You should set the length of this variable with a LENGTH statement prior to calling the routine. If the value of *name* is blank or the routine is being called for the first time in the DATA step, *name* will be replaced by the name of the first variable in the SAS data set. On subsequent calls, *name* is given the name of each variable in the order they are in the SAS data set. After all the names have been returned, VNEXT will return a blank value to *name*.

type is an optional argument. If included, it is the name of the variable that will hold the variable type ('C' = character, 'N' = numeric). Set the length of this variable to 1 before calling the routine.

length is an optional argument. If you include it, you must also supply a variable name to hold the type value. This variable will be a numeric variable and will be the storage length for the variable.

Examples

The variables in these examples are taken from Program 15.1.

Function	Name	Type	Length
CALL VNEXT(NAME, TYPE, LENGTH) 1st call	X	N	8
CALL VNEXT(NAME, TYPE, LENGTH) 2nd call	CHAR	C	1

Program 15.6: Determining the name, type, and length of all variables in a SAS data set

```
***Primary function: CALL VNEXT;

data var_info;
   if 0 then set var;
   length Var_name $ 32 Var_type $ 1;
   do until (missing(Var_name));
      call vnext(Var_name,Var_type,Var_length);
      if not missing(Var_name) then output;
   end;
   keep Var_:;
run;

title "Listing of Data Set VAR_INFO";
proc print data=var_info noobs;
run;
```

Explanation

The trick of using IF 0 (which is never true) allows you to determine variable information without having to actually read any data values from the data set. When the SET statement is compiled, the data set is opened and the variables in the data set are added to the DATA step PDV. During execution, the SET statement never executes. However, the PDV contains all the variables of the data set, so we can now enumerate all of them with CALL VNEXT.

Notice that VAR_NAME and VAR_TYPE are both assigned lengths prior to calling VNEXT. Since VNEXT returns a blank value to the variable named in the first argument when there are no more variables to be read, it is used as a test to determine when to stop execution of the DO loop. Here is the listing of VAR_INFO:

```
Listing of Data Set VAR_INFO

                          Var_
Var_name      Var_type    length

Char          C           1
y             N           8
z             N           8
Money         N           8
x             N           8
a1            N           8
a2            N           8
a3            N           8
Date          N           8
Var_name      C           32
Var_type      C           1
Var_length    N           8
_ERROR_       N           8
_N_           N           8
```

Function: VVALUE

Purpose: To obtain the formatted value of a variable. There is some similarity
between this function and the PUT function. For example, if you have
associated a format with a variable in the DATA step, the VVALUE
function will return a value similar to the PUT function where the second
argument in the PUT function is the format in question. If you have not
previously defined a length for the resulting variable, it will be given a
length of 200. In contrast, the length of a variable created with the PUT
function will be the longest formatted value.

Syntax: VVALUE (*variable*)

variable is the name of a character or numeric variable.

Examples

For this example, the value of Gender is 'M' and it is formatted as "Male".
Date is May 1, 2009 and it is formatted with the DATE9. format.
The value of x is 23 and it is not formatted.

Function	Returns
VVALUE (Gender)	"Male"
VVALUE (Date)	"01May2009"
VVALUE (x)	"23"

Note that the length of each of the returned values will be 200 unless a length was previously
assigned.

Program 15.7: Demonstrating the VVALUE function

```
***Primary functions: VVALUE, PUT;

proc format;
   value $gender 'M'='Male'
                 'F'='Famale';
   value age low-20='Group 1'
             21-30 = 'Group 2';
run;
```

```
data test_vvalue;
   input gender : $1. age date : mmddyy10.;
   length V_Date $ 9 V_Age $ 8;
   format Gender $gender. Age age. Date date9.;
   V_Gender = vvalue(Gender);
   ***the storage length of V_Gender is 200;
   V_Age = vvalue(Age);
   V_Date = vvalue(Date);
   Put_Gender = put(Gender,$gender.);
   ***the storage length of Put_Gender is 6;
datalines;
M 23 10/21/1946
F 77 11/23/2009
x 2 9/16/2001
;
title "Listing of TEST_VVALUE";
proc print data=test_vvalue noobs;
run;
```

Explanation

GENDER is formatted with the $GENDER format. AGE is formatted with the AGE format.
Notice that the results are always character values and the length is 200 unless the length is
previously defined. If you compare the value of V_Gender and Put_Gender, they are the
same except for the storage lengths.

Here is the output:

```
Listing of TEST_VVALUE


                                                          Put_
gender     age        date     V_Date     V_Age    V_Gender  Gender

Male     Group 2   21OCT1946  21OCT1946   Group 2    Male     Male
Famale        77   23NOV2009  23NOV2009        77    Famale   Famale
x        Group 1   16SEP2001  16SEP2001   Group 1    x        x
```

Functions That Determine Format Information

This set of related functions also has corresponding VnameX functions (VFORMATX,
VFORMATNX, etc.) They all return information concerning formats.

Function: **VFORMAT**

Purpose: To return the format name associated with a variable. The name includes the $ if the format is associated with a character variable, and the width, if included. For numeric variables, the name also includes the value to the right of the decimal point. For example, possible returned values are $CHAR15 and DOLLAR8.2.

Syntax: VFORMAT (*variable*)

variable is a SAS variable name.

Examples

The variables in these examples are taken from Program 15.1.

Function	Returns
VFORMAT (CHAR)	$CHAR.
VFORMAT (MONEY)	DOLLAR8.2

See Program 15.8 for an example.

Function: **VFORMATD**

Purpose: To return the format decimal value associated with a variable.

Syntax: VFORMATD (*variable*)

variable is a SAS variable name.

Examples

The variables in these examples are taken from Program 15.1.

Function	Returns
VFORMATD (Z)	4
VFORMATD (MONEY)	2

See Program 15.8 for an example.

Function: **VFORMATN**

Purpose: To return the format name associated with a variable. The name includes the $ if the format is a character variable, but not the width of the decimal value.

Syntax: VFORMATN (*variable*)

variable is a SAS variable name.

Examples

The variables in these examples are taken from Program 15.1.

Function	Returns
VFORMATN (CHAR)	$CHAR
VFORMATN (MONEY)	DOLLAR

See Program 15.8 for an example.

Function: **VFORMATW**

Purpose: To return the format width associated with a variable.

Syntax: VFORMATW (*variable*)

variable is a SAS variable name.

Examples

The variables in these examples are taken from Program 15.1.

Function	Returns
VFORMATW (CHAR)	1
VFORMATW (MONEY)	8

See Program 15.8 for an example.

Functions That Determine Informat Information

This set of functions is similar to the VFORMAT functions described in the previous section except that they return informat information.

Function: **VINFORMAT**

Purpose: To return the informat name associated with a variable. The name includes the $ if the informat is associated with a character variable, and the width, if included. For numeric variables, the name also includes the value to the right of the decimal point. For example, possible returned values are $CHAR15 and DOLLAR8.2.

Syntax: VINFORMAT(*variable*)

variable is a SAS variable name.

Examples

The variables in these examples are taken from Program 15.1.

Function	Returns
VINFORMAT(Y)	READ2.
VINFORMAT(MONEY)	DOLLAR5.
VINFORMAT(CHAR)	$CHAR1.

See Program 15.8 for an example.

Function: **VINFORMATD**

Purpose: To return the informat decimal value associated with a variable.

Syntax: VINFORMATD(*variable*)

variable is a SAS variable name.

Examples

The variables in these examples are taken from Program 15.1.

Function	Returns
VINFORMATD(Z)	2
VINFORMATD(MONEY)	0

See Program 15.8 for an example.

Function: **VINFORMATN**

Purpose: To return the informat name associated with a variable. The name includes the $ if the informat is associated with a character variable, but not the width of the decimal value.

Syntax: VINFORMATN(*variable*)

variable is a SAS variable name.

Examples

The variables in these examples are taken from Program 15.1.

Function	Returns
VINFORMATN(Y)	READ
VINFORMATN(MONEY)	DOLLAR

See Program 15.8 for an example.

Function: **VINFORMATW**

Purpose: To return the informat width associated with a variable.

Syntax: **VINFORMATW(*variable*)**

variable is a SAS variable name.

Examples

The variables in these examples are taken from Program 15.1.

Function	Returns
VINFORMATW(Z)	3
VINFORMATW(MONEY)	5

See Program 15.8 for an example.

Program 15.8: Demonstrating a variety of SAS V functions

```
***Primary functions: VFORMAT, VFORMATD, VFORMATN, VFORMATW, VINFORMAT,
***VINFORMATD, VINFORMATN, VINFORMATW, VLABEL;

data vfunc;
   set var(obs=1);
   length Vformat_x Vformat_y Vformat_char Vformat_money
          Vformatn_x Vformatn_y Vformatn_char Vformatn_money $8.
          Vinformat_x Vinformat_y Vinformat_char Vinformat_money
          Vinformatn_x Vinformatn_y Vinformatn_char Vinformatn_money $8.
          Vlabel_x Vlabel_y Vlabel_char Vlabel_money $ 40;

   ***format functions;
   Vformat_x = vformat(x);
   Vformat_y = vformat(y);
   Vformat_char = vformat(Char);
   Vformat_money = vformat(Money);
   Vformatn_x = vformatn(x);
   Vformatn_y = vformatn(y);
   Vformatn_char = vformatn(Char);
   Vformatn_money = vformatn(Money);
   Vformatw_x = vformatw(x);
   Vformatw_y = vformatw(y);
   Vformatw_char = vformatw(char);
   Vformatw_money = vformatw(Money);
   Vformatd_x = vformatd(x);
   Vformatd_y = vformatd(y);
```

```
   Vformatd_char = vformatd(Char);
   Vformatd_money = vformatd(Money);

   ***informat functions;
   Vinformat_x = vinformat(x);
   Vinformat_y = vinformat(y);
   Vinformat_char = vinformat(Char);
   Vinformat_money = vinformat(Money);
   Vinformatn_x = vinformatn(x);
   Vinformatn_y = vinformatn(y);
   Vinformatn_char = vinformatn(Char);
   Vinformatn_money = vinformatn(Money);
   Vinformatw_x = vinformatw(x);
   Vinformatw_y = vinformatw(y);
   Vinformatw_char = vinformatw(Char);
   Vinformatw_money = vinformatw(Money);
   Vinformatd_x = vinformatd(x);
   Vinformatd_y = vinformatd(y);
   Vinformatd_char = vinformatd(Char);
   Vinformatd_money = vinformatd(Money);

   ***Label information;
   Vlabel_x = vlabel(x);
   Vlabel_y = vlabel(y);
   Vlabel_char = vlabel(Char);
   Vlabel_money = vlabel(Money);
run;

title "Listing of Data Set VFUNC";
proc print data=vfunc noobs heading=h;
   var V:;
run;
```

Explanation

This program is straightforward. Each of the V functions is used on four of the variables from the data set VAR. The informat name for the numeric variables is F. except where an explicit INFORMAT statement is used. The decimal value for the character variable is 0. Inspection of the following listing will help you to clarify this group of V functions.

```
Listing of Data Set VFUNC

Vformat_   Vformat_   Vformat_   Vformat_   Vformatn_   Vformatn_   Vformatn_
   x          y         char       money        x           y          char

YESNO3.    BEST12.    $1.       DOLLAR8.     YESNO       BEST          $

Vformatn_  Vinformat_  Vinformat_  Vinformat_  Vinformat_  Vinformatn_
  money        x           y          char       money        x

 DOLLAR       F.       READ2.     $CHAR1.    DOLLAR5.        F

Vinformatn_ Vinformatn_ Vinformatn_                          Vlabel_
    y          char       money     Vlabel_x    Vlabel_y      char

  READ        $CHAR      DOLLAR   The x variable Pain scale  Char

             Vformatw_  Vformatw_  Vformatw_  Vformatw_  Vformatd_  Vformatd_
Vlabel_money    x          y         char       money       x          y

Housing cost    3         12         1          8          0          0

Vformatd_  Vformatd_  Vinformatw_  Vinformatw_  Vinformatw_  Vinformatw_
  char      money        x            y            char        money

   0          2          0            2            1            5

Vinformatd_   Vinformatd_   Vinformatd_   Vinformatd_
    x             y            char          money

    0             0            0             0
```

C h a p t e r   1 6

Bitwise Logical Functions

Introduction

I doubt you will find any routine uses for these bit manipulation functions in your everyday programming. However, they are fun, and I thought a short program demonstrating the four logical functions (AND, NOT, OR, exclusive OR) would be interesting. I left out the remaining two functions, which shift bits right or left. One reason I decided to include these four bitwise functions is that the exclusive OR function (BXOR) has applications in ciphers and I have long been interested in codes and ciphers.

This collection of functions performs logical operations on bit strings. As a quick review, recall that A AND B is true if both A and B are true; NOT A reverses the value (0 to 1 or 1 to 0); A OR B is true if either A or B is true or if both A and B are true; A XOR (exclusive OR) B is true if A or B, but not both are true. The following table summarizes these rules:

Operation	A	B	Result
NOT A	1		0
	0		1
A AND B	0	0	0
	0	1	0
	1	0	0
	1	1	1
A OR B	0	0	0
	0	1	1
	1	0	1
	1	1	1
A XOR B	0	0	0
	0	1	1
	1	0	1
	1	1	0

A program demonstrating these functions follows the descriptions.

Function: BNOT

Purpose: To negate an argument, i.e., 1's are turned into 0's and vice versa.

Syntax: BNOT (*numeric-value*)

numeric-value is a numeric value that is 0 or positive and not missing.

Examples

For these examples, X = 5 (0101 in binary) and Y = 12 (1100 in binary).

Function	Returns
BNOT(X)	1010 right-most 4 bits in binary, remaining bits are 1's. (Equal to 4294967290)
BNOT(Y)	0011 right-most 4 bits in binary, remaining bits are 1's (Equal to 4294967283)

Function: BAND

Purpose: To perform a logical AND function between two bit strings. As a review, recall that the AND function between two logical variables is true (1) only when both arguments are true. It is false (0) otherwise.

Syntax: **BAND (*numeric-value-one, numeric-value-two*)**

numeric-value-one is a numeric value that is zero or positive and is not missing.

numeric-value-two has the same properties as *numeric-value-one*.

Examples

For these examples, X = 5 (0101 in binary) and Y = 12 (1100 in binary).

```
        X = 0 1 0 1 =  5
        Y = 1 1 0 0 = 12
 BAND(X,Y) = 0 1 0 0 =  4
```

Function	Returns
BAND(X,Y)	4 (0100 in binary)

Function: BOR

Purpose: To perform a logical OR function between two-bit strings. As a review, recall that the OR function between two logical variables is true (1) if either one or both arguments are true. It is false (0) otherwise.

Syntax: BOR(*numeric-value-one, numeric-value-two*)

numeric-value-one is a numeric value that is 0 or positive and is not missing.

numeric-value-two has the same properties as *numeric-value-one*.

Examples

For these examples, $X = 5$ (0101 in binary) and $Y = 12$ (1100 in binary).

```
        X = 0 1 0 1 =  5
        Y = 1 1 0 0 = 12
  BOR(X,Y) = 1 1 0 1 = 13
```

Function	Returns
BOR(X,Y)	13 (1101 in binary)

Function: BXOR

Purpose: To perform a logical exclusive OR function between two bit strings. As a review, recall that the exclusive OR function between two logical variables is true (1) if one (**but not both**) of the arguments is true. It is false (0) otherwise. (Actually, an exclusive OR is just an OR that resides in an upscale neighborhood.)

Note: There are some interesting applications to encoding and decoding text messages using the exclusive OR function. These will be discussed following the encryption and decryption program.

Syntax: BXOR(*numeric-value-one, numeric-value-two*)

numeric-value-one is a numeric value that is 0 or positive and is not missing.

numeric-value-two has the same properties as *numeric-value-one*.

Examples

For these examples, X = 5 (0101 in binary) and Y = 12 (1100 in binary).

```
      X = 0 1 0 1 =  5
      Y = 1 1 0 0 = 12
BXOR(X,Y) = 0 0 0 1 =  1
```

Function	Returns
BXOR(X,Y)	9 (1001 in binary)

Program 16.1: Demonstrating the bitwise logical functions

```
***Primary functions: BAND, BNOT, BOR, BXOR;

title "Demonstrating the Bitwise Logical Functions";
data _null_;
   file print;
   input @1  x binary4. /
         @1  y binary4. /
         @1 Afraid binary8.;

   And = band(x,y);
   Not = bnot(Afraid); ***get it, be not afraid?;
   Or  = bor(x,y);
   Xor = bxor(x,y);
   format x y And Or Xor binary4.
          Afraid Not binary8.;
   put x= y= Afraid= / 60*'-' //
       And= Or= Xor= Not=;
datalines;
0101
1100
11110000
;
```

Explanation

Although the bitwise logical functions work on any numerical arguments, it makes more sense when you either read or output the numbers as binary strings. In this program, the values of X and Y are read in using the BINARY4. informat, while the variable AFRAID is read using a BINARY8. informat. Likewise, the values produced by the functions are also output using BINARY formats. Here is a listing of the output:

```
Demonstrating the Bitwise Logical Functions
x=0101 y=1100 Afraid=11110000
-----------------------------------------------------------

And=0100 Or=1101 Xor=1001 Not=00001111
```

Program 16.2: Enciphering and deciphering text using a key

```
***Primary functions: BXOR, RANK, BYTE
***Other functions: SUBSTR (used on both sides of the equal sign), DIM;

data encode;
   array l[5] $ 1;
   array num[5];
   array xor[5];
   retain Key 173;
   input String $char5.;
   do i = 1 to dim(l);
      l[i] = substr(String,i,1);
      num[i] = rank(l[i]);
      xor[i] = bxor(num[i],Key);
   end;
   keep Xor1-Xor5;
datalines;
ABCDE
Help
;
title "Encoded Message";
proc print data=encode noobs;
   var Xor1-Xor5;
run;

data decode;
   array l[5] $ 1;
   array num[5];
   array xor[5];
   retain Key 173;
   length String $ 5;
   set encode;
   do i = 1 to dim(l);
```

```
      num[i] = bxor(xor[i],Key);
      l[i] = byte(num[i]);
      substr(String,i,1) = l[i];
   end;
   drop i;
run;

title "Decoding output";
proc print data=decode noobs;
   var String;
run;
```

Explanation

This program is an interesting application of the exclusive OR function to the science of cryptography. First, let me explain why the exclusive OR is so useful here. Suppose you have a binary string 11010000 and a key 10101010. If you use an exclusive OR between these two strings, you get: 01111010. What is interesting about this is that you can **reverse** the process by using the exclusive OR again with the encoded string and the key. Thus, the exclusive OR of the result 01111010 and the key 10101010 is the original binary string, 11010000.

Since this is only a demonstration program, the length of the text string was set to five. Obviously, you can expand this to anything you like. You can also use a longer key if you choose. By the way, the RETAIN statement is used both to keep the KEY value in the PDV (program data vector) for the duration of the program and also to set its initial (and only) value to 173.

Before you start to turn your text into code, you use the SUBSTR function to place each of the letters in the original string into the elements of an array, one character to each variable. Since you want to turn text into code and the bitwise logical functions work only on numerical arguments, you use the RANK function to associate a number (the ASCII or EBCDIC collating sequence) with each of the letters. For example, an uppercase A is equal to 65, and uppercase B to 66, and so forth. The BXOR function is used to perform an exclusive OR with each of the numerical values and the KEY. So, if you want to be a spy, you can send the ENCODE data set to someone and make sure the recipient has the key. The encoded text looks like this:

Encoded Message

xor1	xor2	xor3	xor4	xor5
236	239	238	233	232
229	200	193	221	141

The decoding section of the program is similar to the encoding section. The exclusive OR is performed on every coded value and the key. The resulting letters are the original text, and they are placed in the proper position in the new string using the SUBSTR function on the left-hand side of the equal sign. This use (see Chapter 1) allows you to place individual characters in any position of a character variable. The decoded text is turned back to the original message and is shown here:

```
Decoding output

String

ABCDE
Help
```

Encrypting and Decrypting Macros

The two macros that follow are general purpose encrypting and decrypting macros that make use of the exclusive OR function (BXOR) and the uniform random number function. Rather than use a single key to perform the encryption, the key is actually a random number seed which, in turn, generates a whole series of keys, one for each character in the document to be encrypted. It turns out that if you have a truly random key the same length as the document you want to encrypt, it is theoretically unbreakable! In the spy business, high-security messages can be transmitted using a "one-time pad" (or so I'm told). The problem with this system is that the recipient of the coded message needs to keep a copy of the keys somewhere, probably in a safe, and it could be stolen or somehow compromised.

The macros presented here rely on the SAS random number function (RANUNI) to generate a random, but repeatable, sequence of keys. All that you need to reproduce the series of random numbers is the seed value. So, this makes for a fairly secure code if you choose a key with a large number of digits (say 9 or 10). Following the two macros, there will be a fairly detailed explanation.

Program 16.3: Writing general-purpose encrypting and decrypting macros

```
***Primary functions: BXOR, RANK, BYTE
***Other functions: SUBSTR (used on the left-hand side of the equal
sign), DIM, RANUNI;

%macro encode(Dsn=,        /* Name of the SAS data set to hold the
                              encrypted message */
           File_name=,   /* The name of the raw data file that holds
                              the plain text */
```

```
            Key=             /* A number of your choice which will be the
                                seed for the random number generator. A
                                large number is preferable */
            );
   %let len = 80;
   data &dsn;
      array l[&len] $ 1 _temporary_; /* each element holds a character
                                         of plain text */
      array num[&len] _temporary_;   /* a numerical equivalent for each
                                         letter */
      array xor[&len];               /* the coded value of each letter */
      retain key &key;
      infile "&file_name" pad;
      input string $char&len..;
      do i = 1 to dim(l);
         l[i] = substr(string,i,1);
         num[i] = rank(l[i]);
         xor[i] = bxor(num[i],ranuni(key));
      end;
      keep xor1-xor&len;
   run;
%mend encode;

%macro decode(Dsn=,        /* Name of the SAS data set to hold the
                              encrypted message */
            Key=           /* A number that must match the key of
                              the enciphered message */
            );
   %let Len = 80;
   data decode;
      array l[&Len] $ 1 _temporary_;
      array num[&Len] _temporary_;
      array xor[&Len];
      retain Key &Key;
      length String $ &Len;
      set &Dsn;
      do i = 1 to dim(l);
         num[i] = bxor(xor[i],ranuni(Key));
         l[i] = byte(num[i]);
         substr(String,i,1) = l[i];
      end;
      drop i;
   run;
   title "Decoding Output";
   proc print data=decode noobs;
      var String;
   run;
%mend decode;
```

Here is a sample calling sequence:

```
%encode (Dsn=code, File_name=c:\books\functions\plaintext.txt,
key=17614353)
%decode (Dsn=code, Key=17614353)
```

Explanation

The calling arguments are the name of the output SAS data set that will hold the coded message; the full path and filename of the raw, plain text message; and a key value (which is really the seed for the random number function).

To make the program easier to adapt, I assigned the maximum length of each line of text to a macro variable (LEN). If you want, you could just replace the &LEN values in the macro with a fixed value, such as 80.

The L array stores each of the individual characters in the plain text message. The NUM array assigns a numerical equivalent to each of the character values, using the RANK function. The BXOR function performs an exclusive OR between each of the numerical equivalents and a random number generated using the RANUNI function (and seeded with the KEY value). Since you don't need any of the L or NUM variables in the data set, these arrays are defined as _TEMPORARY_ to make the program more efficient.

The resulting data set is what you can transmit to your spy friends. Hopefully, you have arranged for him or her to know the value of the key. Perhaps the key value can be computed from the date in some manner.

The deciphering macro just reverses the process. Here, you use the same seed for the random number function, so you generate the same series of random numbers. I suspect that the random series may be system dependent, so you may want to be sure that you and your spy friends are both on PCs or mainframes, etc. Also, you need to be sure that both systems use the same coding system (ASCII or EBCDIC).

Once the DECODE macro has recomputed the numerical values for each letter, the BYTE function (the inverse of the RANK function) turns these numbers back into the plain text letters. Finally, the SUBSTR function places each of these characters in the proper position in the STRING variable. I have no idea whether the NSA and their supercomputers could break this code easily or not. But, if they didn't run SAS on their machines, they would have lots of problems!

Anyway, I hope you enjoyed this novel use of the BXOR function. I had a lot of fun writing it!

List of Functions

Index

CPSIA information can be obtained at www.ICGtesting.com
Printed in the USA
LVOW020500141112

307168LV00004BB/36/P